W9-BAR-662

The Secret of Life

Redesigning the Living World

Joseph Levine and
David Suzuki

The Secret of Life

Redesigning the Living World

WGBH Boston

Library of Congress Cataloging-in-Publication Data

Levine, Joseph S.
The secret of life: redesigning the living world/
by Joseph Levine and David Suzuki. — 1st ed.
p. cm.
"A companion to the PBS television series."
Includes bibliographical references (p.) and index.
ISBN 0-9636881-0-3
1. Molecular biology—Social aspects.
2. Molecular genetics—Social aspects.
3. Genetic engineering—Social aspects.
4. Biotechnology.
5. DNA.
I. Suzuki, David T., 1936- . II. Secret of life (Television program). III. Title.
QH506.L43 1993
574.87'328-DC20 93-4817 CIP

Design and production by Douglass Scott and Cathleen Damplo

Manufactured in the United States of America
First edition

Contents

Introduction

Viewed from the distance of the moon,
the astonishing thing about the earth, catching the breath,
is that it is alive.

Lewis Thomas

FROM SPACE, THE earth is dazzling, a planet wrapped in a majestic mantle of living organisms. It is also a cosmic paradox. For at every level of organization, from the largest ecosystem to the tiniest cell and the molecules within, the beauty and order of life violate entropy, the tendency of the universe to drift toward disarray. The universe expands, stars exhaust themselves, and energy dissipates in space. Yet across the surface of our planet, entities we describe as being alive bind atoms into complex arrays that replicate, accumulate, and diversify over time.

At the root of these living organisms, these ordered eddies in the river of chaos, lie the storehouses of biological information, molecules of deoxyribonucleic acid—DNA. You can actually see DNA with the naked eye; if you take a handful of cells and break them apart with the right kind of detergent, they release a clear, viscous substance that you can twist like caramel around a glass rod. The nondescript appearance of this extract belies both its power and its history, for this is the substance of heredity, the repository for all the information needed to build a bacterium, a corn plant, a mouse, or a human being.

DNA's power is global; it has orchestrated the history of life on earth for three and a half billion years. Yet its touch is intimate; it determines your chances of getting cancer, the amount of cholesterol in your father's blood, and the color of your daughter's eyes. Discoveries about its actions have sparked a molecular revolution whose consequences affect every corner of both pure and applied biological research.

Like other revolutions, this one has roots that reach deeply into the past. Charles Darwin observed, though he couldn't explain, the existence of heritable variation in wild and domestic animals and plants. He knew, therefore, that living things somehow store, utilize, and transmit to their offspring the biological information which makes them what they are. Gregor Mendel, among the first to study the actions of heredity, called the controlling elements behind traits in his pea plants *merkmallen* after the German word for "characters." In 1942, many blind alleys and false leads later, experiments suggested that DNA was the most likely carrier of hereditary information. By that time, Mendel's *merkmallen* were called *genes*. Then, on April 25, 1953, James Watson and Francis Crick published their paper "A Structure for Deoxyribose Nucleic Acid," in *Nature* magazine.

Watson and Crick's Nobel Prize–winning discovery of DNA's molecular structure precipitated the kind of perceptual shift that is vital to the progress of science. In the past, as studies of geology and astronomy matured, the flat earth first ballooned into a globe, then fell from its place at cosmic center. In the hands of physicists, the indivisible atom shattered first into electrons, protons, and neutrons, then into quarks, leptons, and gluons. And in biology, the nucleus of the living cell began to divulge its secrets.

The revolution in the biological sciences that followed has enabled biologists to view the biosphere from the perspective of inner space, the microworld of molecules that control the processes of life. Explorations of that world have already revolutionized the way all of us interact with other living entities. Researchers can now locate and identify individual genes, the functional units of DNA. They can pluck those genes from one living cell, copy them in a test tube, move them into the cells of another species, and splice them onto those cells' native DNA as deftly as you might splice a new scene into movie film. They can induce those cells to obey their newly acquired genetic instructions when and where they want them to. Cells treated this way, and their progeny, will never be the same. The technical term applied to them is both accurate and descriptive: in the jargon of molecular biologists, these cells have been *transformed*.

Transformation and other techniques of the new biology affect much more than a few cells in laboratory culture dishes. No new body of knowledge since the development of atomic energy has offered such extraordinary and far-reaching opportunities to improve the human condition. That may sound like hyperbole, but it isn't. Genetically engineered organisms from bacteria to cows are already producing insulin, human growth hormone, and other heretofore prohibitively expensive compounds used in treating both infectious and degenerative diseases. The genes of wild animals and plants will soon be providing instructions for the design of new drugs and industrially important compounds, ranging from adhesives to fibers to lubricants. Newly approved applications of genetic testing and human gene therapy are already beginning to alter the way we are born, the way we fight disease, and the way we die. Small wonder that many observers consider this technology among the greatest astonishments of the modern era.

Not surprisingly, the power and potential of molecular biology have made it more controversial, even among its practitioners, than any branch of life science has ever been. The major players in the drama—scientists, politicians, entrepreneurs, physicians, advocates, patients, private corporations, and public institutions—differ wildly in intellectual sophistication and goals, and all relate to the field in their own way.

Some researchers are buoyant optimists and indefatigable boosters of genetic medicine. Others spend much of their time helping to create legal guidelines to prevent abuse in ethical gray areas of the field. Still others dedicate themselves to finding socially responsible uses for new technologies almost before they are out of the test tube. And a cadre of postmodern Luddites, in addition to pinpointing legitimate social concerns, disseminates apocalyptic warnings aimed at rallying those terrified of a technology they don't understand.

Biotechnology companies, often thriving on the contested border between publicly funded and private research, aim to put molecular techniques to useful service while pursuing a handsome profit. Health insurance companies are more circumspect in their response to molecular tools; reeling from skyrocketing medical costs, they see genetic testing as a means to identify, and isolate from general risk pools, unprecedented numbers of high-risk individuals.

What's the average citizen to make of this fascinating and confusing mix of scientific advances, strong personalities, and conflicting claims? Even highly educated laypeople find the science itself abstract, dense, and remote from daily experience. In addition, advances in the field are sometimes subtle, nearly always complex, and generally predate practical applications by several years. That makes a tough recipe for mass-media news stories, which gravitate toward easy, sensational reports—*the* cure for cancer, *the* vaccine for AIDS, *the* gene for alcoholism—rather than the long series of partial discoveries, near misses, and reevaluations that characterize research. Even the finest science writers at the best newspapers shoulder a nearly insurmountable burden; their articles must explain complex, often contentious findings in isolation from the body of scientific thought that informs them. Readers, lacking both the context in which to consider new developments and the background necessary to evaluate conflicting claims, find themselves at sea.

That predicament launched both *The Secret of Life* television series and this book. Back in 1986, Paula Apsell, executive producer of *Nova*, and Graham Chedd, executive producer of this series, decided that the time had come for a large-scale, coordinated effort to introduce PBS viewers to the new biology, its applications, and its social ramifications. Thomas Levenson, then science editor of *Nova*, and Joe Levine, who in addition to being co-author of this book would become science editor for the series, began a six-month whirlwind of meetings, phone calls, and discussions with prominent researchers, *Nova* advisers, science journalists, and friends. We assembled a dedicated series advisory board centered at, but not restricted to, MIT's Whitehead Institute for Biomedical Research.

All four of us knew from the start that the breadth and depth of the science involved precluded a comprehensive review of all the work going on in molecular biology. Our problem was not digging up stories but weeding them out, not finding important science but deciding which discoveries, researchers, and social issues we could meld into a coherent presentation. We knew that the brisk pace of progress in the field would make it impossible for either the series or this book to deliver the final word on any subject. In fact, we look back now with both humor and amazement at the original series proposal, barely seven years old; much of what was then no more than speculation has already been accepted as fact or dismissed as fantasy, while several fantasies of those years have been transformed into the realities of today.

What we've tried to assemble is a progress report grounded in advancing research and leavened with social relevance. We aimed for a mix of people, ideas, advances, and dilemmas that would convey a feeling for how we know what we know and why we don't know things that we don't yet understand. We searched for personal stories that would be current but not trendy, research projects that would be idiosyncratic enough to be intellectually interesting yet sufficiently grounded in fundamental concepts to have important potential applications. Certain broad subject areas—cancer, immunology, genetic engineering, gene therapy—were always givens. Others—the unity and diversity of life, the influence of genes on human behavior—were sufficiently abstract and complicated to give us pause.

Then series host and co-author, geneticist and science educator David Suzuki signed on. Our collaboration with the BBC was solidified under project director Robin Brightwell, and our team of American and British producers came together. Calling again on *Nova*'s friends and advisers, we set up individual meetings, group conferences, and a "molecular biology minicourse" during which prominent researchers addressed the entire team for hours at a time. Ultimately, we settled on the content and order reflected in this book.

The first chapter sets the stage by presenting the most basic insight of the new biology: that all things we describe as being alive—animals, plants, and even bacteria living in volcanic hot springs—are united by shared information stored in DNA's double helix. That biochemical unity offers both challenges and opportunities as researchers struggle to determine how DNA stores and transmits information from generation to generation and how its coded instructions are translated into pairs of wings, iridescent fins, a turned-up nose, or a faulty heart.

Chapter 2 explores a central question in biological thought: If DNA stands at the root of life's unity, how did it also enable the evolution of biodiversity, the dazzling array of between 2 and 40 million fellow organisms with whom we share the earth? Ever since Darwin, biologists have searched for a coherent theory to explain that diversity. How does life evolve? How are new species formed? Why do some evolve into scores of robust decendants while others slip toward extinction? Chapter 2 explains how our newfound ability to read and manipulate genes allows us to test evolutionary theory in ways that Darwin never dreamed of, and to answer questions about the history of life that fossils cannot resolve. This chapter reveals that the double helix is at once the root of evolutionary change, a record of changes past, and a key to understanding—and valuing—diversity today.

Chapter 3 focuses on what many biologists view as the most fundamental question in molecular biology today: How does a single fertilized egg cell divide and change to produce trillions of functioning skin, bone, muscle, and nerve cells, all of which contain the very same genes yet grow into different shapes and perform different tasks? Most of us, if we think of genes at all, think of those which make our eyes blue or give us our father's nose. But genes that control growth and differentiation are different; these "master controls" regulate a host of other genes. One such stretch of DNA, dubbed the "male" gene, may determine one of the most complex aspects of the human condition; a fertilized egg containing this gene grows into a man, while an egg lacking it becomes a woman.

Chapter 4 explores what happens when certain genes that normally control growth and development fail in their work. The result can be summed up in a single word: *cancer.* This chapter explains how molecular biology is revealing that, behind its diverse causes and symptoms, cancer is a disease of the genes. Researchers have learned that carcinogens and heredity can conspire to produce stretches of DNA that override built-in growth controls and transform normal cells into tumor cells—potentially immortal parasites that reproduce out of control, spreading throughout the body and killing their host. The search for treatments to reverse that process—and for tools to detect damaged genes and prevent them from wreaking havoc to begin with—is an ongoing process that affects us all.

Of all the organisms that share the earth with us, the only ones that cause disease are those which are able to enter the body and evade its defenses. Chapter 5 explains that because those invasions are directed by parasites' DNA, their genes and the genes of our immune system are locked into a perpetually escalating arms race; through time and evolutionary change, our body struggles to fight them off, while they battle to survive at our expense. This chapter explains how the genes that guide the cells of our immune system "learn" to distinguish self from nonself, to attack intruders, and to remember infections as defense against recurrent attacks. Thanks to a combination of molecular and medical intelligence, some of our most ancient enemies, such as smallpox and polio, have been virtually vanquished. Battles with others, ranging from schistosomiasis to AIDS, are pushing researchers beyond the limits of current knowledge.

"Ice nine" bacteria that prevent frost damage to strawberries, tomatoes that don't rot in shipment, and genetically engineered growth hormone that increases cows' milk production are just a few tangible products of genetic engineering, the field that dominates Chapter 6. Its potential is astonishing, for it utilizes the power not just to read the genetic code, but to rewrite the chapters that define life around us, literally designing the world to order. This chapter follows the development of tools that enable researchers to manipulate DNA much the same way that a word processor edits manuscripts. It also traces the connection between scientists, experiments, and commercial applications, and shows how misinformation, fear, and distrust encumber already difficult decisions on the use of molecular technology.

Chapter 7 follows genetic engineering as it manipulates genes in that most peculiar of subjects, *Homo sapiens.* We have a special name for this particular application—we call it gene therapy, and it holds potential to alleviate pain and suffering of current and future generations by eliminating, replacing, or correcting damaged genes that cause inherited disease. The chapter also explores the technology's dark side, noting the potential of genetic diagnostics to foster discrimination and pointing out the slippery slope that begins with the frivolous use of technology for cosmetic enhancement.

The final chapter explores molecular biologists' struggle to chart the interactions between the genome and the brain, arguably the two most complex systems in the living world. Past efforts to link genes with mind include many examples of hideously bad science, from the misguided through the racist to the patently deceitful. Yet new data seem to link genes with schizophrenia, alcoholism, risk taking, and other complex behaviors. This chapter attempts to present enough background to pose questions not asked often enough. Given what we do and don't know about DNA, is it reasonable to suggest that genes fine-tune subtle aspects of personality? If so, given ethical constraints on human experimentation, is there any way to prove such a connection and identify the responsible genes?

All of us involved in the book and the series have always realized our duty to present an appealing and accessible balance of the science that underlies the molecular revolution, the intellectual fascination that drives its researchers, and the political realities that confront applications of their work. One of us (J.L.), as a writer and teacher involved in science education at both high school and college levels, and the other (D.S.), as a university academic and broadcaster, are personally and acutely aware that public understanding of the relevant science is at best spotty and at worst nearly nonexistent. Coming of age academically during the 1970s, Joe witnessed firsthand the philosophical and political struggles surrounding the rebirth of interest in the contribution of genes to behavior and the first major successes and failures of the molecular revolution. He is unfortunately familiar with the tragedy of discrimination based on pseudo-scientific notions of racial genetics: parts of his mother's family disappeared in concentration camps. He also knows intimately the medical and social issues surrounding genetic disease, having inherited from his father a gene which causes Familial Hypercholesterolemia. As a third-generation Canadian of Japanese ancestry, David experienced the sting of racism during the Second World War when he and his family were classed as enemy aliens and incarcerated for three years. Only as a faculty member and practicing geneticist did he come to realize that the climate for their incarceration had been fostered by the exuberant claims of geneticists early in this century.

We both realize science has implications that ripple far beyond the boundaries of the scientific community, universities, or corporations. The power of the new biology carries with it the responsibility to use that power wisely. Genetic engineering and biomedicine, through abilities that draw ever closer to "playing God" with DNA, raise awesome ethical, legal, and social questions. What may be the ecological impact of new life forms, and how do we handle their legal status as "property"? What guidelines, if any, should regulate the manipulation of human genes in efforts to cure inherited diseases? How will knowledge gained from genetic testing affect your chances of securing a job or obtaining health insurance?

These questions have no simple answers because they involve far more than just scientific information, and because no well-informed individual—scientist or layperson, physician or patient, employee or employer, cleric or politician—can claim to be a disinterested observer. Molecular biology, through its ability to transform cells in test tubes, has acquired the power to transform the most intimate details of our lives and the broad sweep of social policy. Nearly all its applications either threaten or bolster the health, livelihood, political convictions, social agendas, or religious beliefs of some segment of the population. The subjects we discuss here refuse to stay closeted in laboratories and scientific journals. In order to avoid mistakes of the past, the public has a vital stake in participating in the debate about where science is leading. This becomes especially important as we witness the explosive growth in genetic manipulative capacity and the sometimes outlandish claims of its most ardent practitioners.

That's why we have written this book. In an effort to foster rational debate on issues of incalculable importance, we have tried to convey the effort and the exhilaration, the promises and the dangers, of a scientific revolution in progress.

1

The Immortal Thread

We have entered the cell, the mansion of our birth,
and have started the inventory of acquired wealth.

Albert Claude

"Some people," says German biologist Karl Stetter, chuckling, "call me the King of Hades." As he surveys the smoldering crater of Solfatarra, near Naples, Italy, it's easy to see why. Around him the air is redolent with sulfur, superheated steam erupts from fumaroles, and mineral-laden mud boils out of volcanic vents. Yet Stetter takes a deep breath and grins. "For me," he explains, "the smell of sulfur and brimstone is not hell, not Hades, it's just heaven! I like it because I like these guys which are growing here, and I want to investigate them."

The "guys" Stetter refers to are bacteria thriving under conditions that would kill nearly any other living thing. The belching mud pots around him are near or above boiling, virtually devoid of oxygen, and laced with sulfuric acid. Nonetheless, Stetter is looking for life. "I'm absolutely sure that organisms are there," he explains. "They are living down in the superheated water. Watch your step! This is not a safe area," he warns coworkers as he leads the way, testing the mineral-encrusted ground before him with a long metal pole. Reaching a likely spot, he inverts the pole and dips a small can welded to its other end into an erupting fumarole. "Be careful," he counsels an associate whose gloved hand holds a jar ready to receive the bubbling brew. "It's damn hot!"

Until the 1980s, few biologists believed that anything could live under such extreme conditions. Even Stetter, one of the pioneers who challenged that view, doesn't know how his subjects survive. "One of our research targets is DNA," he explains. "It's a miracle why it's stable at these high temperatures. It should fall apart, but it obviously does not, because these bugs are growing here."

To find out how his bugs survive, Stetter takes samples back to his home base in Regensburg, Germany, to culture them in what he jokingly calls his witches' kitchen. There, rows of stainless steel vats, designed to duplicate the conditions under which the microbes grow, incubate them at high temperatures, re-create their oxygen-free environment, and feed them their unlikely food of choice, sulfur. Periodically, Stetter's team harvests the crop, dismembers the bacterial cells, and analyzes their contents. Their studies have turned up several biochemical surprises, including one that even Stetter marvels at: these microbes, which are poisoned by the oxygen we breathe, live in water that would boil us alive, and eat sulfur, contain stretches of DNA that strongly resemble certain human genes. In biological terms, some of the information that keeps them alive looks remarkably like the instructions for life that we carry around inside our own cells.

What's so extraordinary about this? After all, biologists expect similarities among organisms. Evolutionary theory teaches that all living things descend from common ancestors, and with relatedness comes resemblance. But as relationships grow more distant, the degree of expected similarity decreases. You look a lot like your parents, somewhat less like a chimpanzee, and much less like a rabbit. Still, chimps and rabbits, with whom humans share more characteristics and more DNA than you might expect, are both mammals, cousins, so to speak, whose forebears diverged from ours no more than 200 million years ago—not all that long on the scale of evolutionary time. Sulfur-eating bacteria, by contrast, which last shared common ancestors with humans more than 3.5 *billion* years ago, differ from us in form and habit more than almost any other living thing. Yet profound similarities between our genes—the biological instructions that keep us alive—remain.

"I am brother to dragons," reflects the Book of Job (30:29), and that's precisely what the first stages of the molecular revolution have driven home to researchers. Stetter's work is hardly unique; a steadily growing body of similar findings continues to surprise researchers with the extent of life's unity at the molecular level. Biologists, who have long taken for granted a theoretical understanding that all life is connected, are faced with profound biochemical evidence that connections among organisms are tangible, functional, and *alive.*

"There isn't a single scientist I know who isn't amazed," admits Gerry Fink, an MIT biomedical researcher whose work leads him daily to compare human genes with yeast genes. "We all knew that evolution was true, but now, every time I pick up a cell, I have the same amazement. These genes really are there, and they are the same genes across species. A little bit of tinkering here and there, that's all. We really are connected to all these organisms."

To appreciate fully why these discoveries are reshaping the biological sciences, and why they have profound implications for medical research and our broader understanding of the living world, we must descend from the familiar domain of visible life into the unseen microcosmos of cells and molecules. Any living plant or animal—pine tree or parrot, primrose or person—is a collection of millions, even trillions of working units called cells. Not all cells in an organism are alike; hundreds of different types perform radically different tasks. Some grow hair while others build teeth; some produce milk while others secrete digestive enzymes. Yet all those cells coordinate their actions precisely to function as a single entity. How do they do it?

The information that guides those cells resides in their DNA. If you could peer into a typical cell, you would see a compartment called a nucleus near its center. Within each nucleus are chromosomes, tangled tufts that contain DNA's long, twisted strands. Those strands of DNA, in turn, carry all the biochemical instructions necessary to produce and coordinate the components of living tissue.

That molecular intelligence is awesome, in both the classical and the colloquial senses of the word. Its list of instructions is incredibly long; if you could unravel all the DNA in your body and stretch it out end to end, it would reach to the sun and back more than a hundred times. The information stored in all that DNA controls more details of your life than you have probably ever imagined, from obscure subcellular processes to a host of characteristics that make you the individual you are. Some genes interact with the food you eat and the environment you live in to determine your height, the amount of fat and muscle you carry, and your risk of getting a heart attack. Not only do certain genes govern the color of your hair, but others determine where on your body that hair will grow, when it will start growing, and how long it will continue to grow. If, as some of us approach middle age, we find more hair on our back and less on our head, genes are at the root of it all.

In addition, as Stetter, Fink, and their colleagues have discovered, both DNA's structure and many of its messages are common to every living being on earth. If researchers had the skill, they could remove certain lengths of DNA from human cells, insert them into yeast, and replace them with fragments from the cells of a giant squid, all without making any difference in the lives of those organisms. In a real sense, therefore, the entire biosphere—from Stetter's bacteria to Fink's yeasts and humans—is animated and governed by various forms of this unitary molecular intelligence.

The common molecular language of life also underlies infectious disease. Viruses can give you influenza or AIDS because they speak the same genetic language that your cells do. Given the opportunity to infect you, they interact directly with your genes, subverting your molecular machinery to their purposes—and your peril. On a different level, battles between other disease-causing organisms and your immune system pit their genes directly against yours in recurrent episodes of molecular espionage and counterespionage.

The power of genes, combined with the extent to which we share genetic information with other living things, has enormous practical and philosophical significance. For example, a stretch of human DNA that could give you cancer if it malfunctions also helps control how yeast cells divide. That's why Gerry Fink and his colleagues can learn about human cancer by studying the genes of yeasts—not because those organisms resemble us superficially, but because they work the same way we do at the molecular level.

This insight into life's biochemical fraternity carries with it a dose of philosophical irony. For the science that spawned it matured in Western culture, which teaches that humans are distinct from the rest of the living world, and that we enjoy divinely granted dominion over the globe. We "conquer" frontiers, "tame" the wilderness, and "control" the forces of nature. Our science is reductionist; to understand complex systems we dismantle them into smaller parts, search for rules that govern those parts, and use those rules to predict actions of the whole.

Many other cultures, by contrast, view humans as but one member of a community that includes all living things and is governed by an omnipresent power or "way." The laws of nature conceived within such world views are often holistic, based on observations of interacting components rather than on suppositions of what those components might do if isolated from the whole. Common to most such styles of thinking is the attitude that humans must recognize and respect the ties that bind us to the mortal and spiritual dimensions of other living things.

You may already have noticed the irony: by looking at ever-smaller units of living systems, our science has found stretches of DNA that our species shares not only with animals like wolves and eagles that indigenous traditions recognize as our brothers and sisters, but with Fink's yeasts and Stetter's bacteria. Such shared bits of molecular intelligence are physical proof that our common biological heritage unites us all under the same basic laws of life. Non-Western world views, in a sense, recognize that heritage and extend it from the strictly physical to the spiritual level.

What are the practical implications of this unifying perspective? Before we can explore them in detail, or even explain exactly where the common molecular heritage comes from, we must answer several basic questions: Exactly what *is* DNA? How did we learn about it? What does it look like? How does it carry information? and What does that information actually do?

• • •

TEXTBOOKS USUALLY HONOR Gregor Mendel, an Austrian-born priest who experimented with peas in a Czechoslovakian monastery, as "the father of heredity." But more than twelve hundred years before Mendel counted pea pods, a passage in the Talmud advised that a male child whose maternal uncle was a "bleeder" should not be circumcised, for he might bleed as well. Farmers have used seat-of-the-pants reckoning to breed desirable traits into plants and animals since the dawn of civilization. Still, many people believed that heritable traits somehow "blended" between generations; crosses between large dogs and small dogs produced medium-size dogs, after all, and mixtures of parental traits among human children were obvious.

Mendel proved otherwise, showing that although both parents contribute to their offsprings' features, their traits don't blend. He proposed that heritable characteristics are controlled by units of inheritance we now call genes. Presciently, he suggested that each plant carries two genes for every character trait and that particular genes have different forms. The gene for flower color could have, for example, both a white form and a purple form. We now call such forms *alleles*. Finally, in a stroke of genius, Mendel proposed that plants' reproductive cells each carry only one allele for each gene.

In 1865, Mendel presented to the Society of Natural Science in Brünn his ground-breaking work, *Experiments in Plant Hybridization*. It was politely received and virtually ignored until the beginning of the twentieth century. Three years later, Johann Meischer, a Swiss researcher, began studying the chemistry of the cell nucleus. In 1869 he discovered a mixture he called *nuclein*, which contained large molecules mixed in with smaller proteins. By 1870, working with an extract from the sperm of salmon caught in the Rhine, Meischer purified the larger component, which one of his students later named *nucleic acid*. Meischer, however, didn't realize the significance of his discovery. He believed that large molecules could carry complex messages in repeating patterns of smaller components. He even suggested that biological information could be encoded in much the same way that "words and concepts of all languages can find expression in twenty-four to thirty letters of the alphabet."[1] But he, his colleagues, and many of their successors didn't think of nucleic acid in that context. As geneticist Thomas Lee of Saint Anselm College observed, "The bottle of innocent white powder was put away on the laboratory shelf. It would be sixty years before it would be revealed to be, in fact, what it was, a bottle of genes."[2]

Within a few decades, researchers learned that a particular form of nucleic acid, specifically deoxyribonucleic acid, or DNA, was the main constituent of chromosomes, but they still didn't recognize its importance. In fact, DNA seemed so chemically uninteresting that, by association, the chromosomes it formed were thought to be mere physical supports, internal stiffeners of no more importance to the cell than, as Horace Judson put it, "the laundry cardboard in the shirt, the wooden stretcher behind the Rembrandt."[3]

Slowly and erratically over time, insights from varied fields began to implicate DNA as an important player in the life of the cell. Around the turn of the century, geneticists rediscovered Mendel's work and began to compare notes with colleagues who were examining cells through new and better microscopes. Studying organisms as diverse as primroses, fruit flies, and molds, they discovered that genes behaved—first suspiciously, then undeniably—as though they were located on chromosomes.

Then in 1928, several baffling experiments by Frederick Griffith, a British Public Health Service scientist, pointed the way toward the carrier of heredity. Griffith worked with two strains of *Pneumococcus* bacteria, labeled *S* (for smooth) and *R* (for rough) after the appearance of their colonies in culture. The two strains differed in an important way. When injected into mice, living bacteria from the S strain caused pneumonia. Injections of dead bacteria from that same strain, not too surprisingly, did not cause disease. By contrast, bacteria from the R strain, whether alive or dead, didn't produce pneumonia. But then Griffith uncovered a puzzling phenomenon: when he injected dead, disease-causing bacteria *together with* live, harmless bacteria, the mice developed pneumonia!

This was astonishing. Neither injection alone caused disease, so how could their combination be deadly? What's more, when Griffith isolated live bacteria from the sick mice, those microbes grew into colonies that resembled the original disease-causing strain and permanently retained their ability to produce pneumonia. Apparently some component of the dead, disease-causing bacteria had transformed individual microbes from the formerly harmless strain. What came to be known as the *transforming principle* had somehow caused a change that was passed on from generation to generation.

Many scientists refused to believe Griffith's results, even though they were replicated and extended by independent research teams. One researcher willing to give Griffith the benefit of the doubt was Oswald Avery of New York's Rockefeller Institute, who, in 1944, determined to identify this mysterious transforming principle. Avery's team took an extract from Griffith's original S-strain, disease-causing bacteria and treated it so as to destroy every type of compound then thought to carry genetic information. The extract still transformed the R strain as it had done in Griffith's hands, but when Avery destroyed DNA in the extract, transformation no longer occurred. The evidence seemed incontestable: the transforming principle must be composed of DNA.

For a time, however, many biochemists insisted that DNA was too "stupid" a molecule to carry much information. It contained, after all, only a rather boring sort of sugar, a group of oxygen atoms clustered around a phosphorus atom, and four rather similar bases—the compounds *adenine, guanine, thymine,* and *cytosine.* Those components seemed too simple and repetitive to carry information—or were they? Perhaps something unique about DNA's structure could solve the puzzle.

Erratically, clues about that structure which would ultimately clinch DNA's role fell into place. During the late 1940s, biochemist Erwin Chargaff, then of Columbia University, noticed that the relative proportions of DNA's four bases do not vary independently of one another. Instead, the percentages of guanine and cytosine are always nearly equal, as are the percentages of adenine and thymine. Any structure proposed for DNA had to account for this pattern.

Meanwhile, Rosalind Franklin, a talented researcher working at King's College during the early 1950s, provided another vital piece of information. Franklin, forcing a suspension of DNA molecules to line up parallel to one another in a thin glass tube, passed X rays through them. The scattered pattern the X rays produced on a photographic plate could be interpreted by employing the insights of X-ray crystallography. Franklin's work provided three vital clues to DNA's structure. First, some parts of the molecule were arranged like the rungs of a ladder. Second, certain structures repeated at specific, regular intervals. Third, the molecule probably had some sort of twisting, helical shape that also repeated at regular intervals. Still, no one could assemble all the scattered clues until two young scientists at the nearby Cavendish Laboratory of Cambridge University learned about Franklin's data.

James Watson and Francis Crick were outsiders to biochemistry. Crick, like many early molecular biologists, came to the field from physics; Watson had worked with viruses that infected bacteria. Most biochemists of that day were interested in enzymes, energy transfer, and chemical reactions. But Watson and Crick were interested in information and structure. They had been working with molecular models, twisting bits of wood and paper into various shapes to help them visualize how DNA's parts could form a structure that could do everything DNA was able to do.

According to his account in *The Double Helix,* Watson, after hours of theorizing and modeling, caught a glimpse of Franklin's X-ray pattern. "The instant I saw the picture," Watson wrote, "my mouth fell open and my pulse began to race."[4] Soon thereafter, he and Crick proposed the double helix model in *Nature* magazine. Their work earned them the Nobel Prize for physiology or medicine in 1962.[5] Molecular biology was off and running.

Watson and Crick's model for the structure of DNA speedily gained wide acceptance because it accomplished two vital tasks. First, it accounted for all the available chemical and physical evidence. Second, it led the way toward more detailed explanations of how DNA performs the functions necessary for it to serve as the carrier of hereditary information.

THE WATSON-CRICK MODEL is quite simple, considering all that DNA must do. Each of the molecule's basic units, called *nucleotides,* is composed of three simple parts that had been recognized for years: the sugar deoxyribose, one "arm" consisting of a few oxygen atoms clustered around a phosphorus atom, and another arm consisting of adenine, guanine, thymine, or cytosine. The interesting feature of these nucleotides is that when they link up to one another chemically, their sugars and phosphorus groups form a long, single-stranded chain, and their bases stick out sideways.

After looking at Franklin's data, Watson and Crick realized that two such chains could line up in parallel, with their bases facing each other. Watson and Crick also suggested that the chemical properties of those bases would always cause adenine (A) to pair with thymine (T), and guanine (G) to pair with cytosine (C), rather like the way the symbols for yin and yang fit together. Calling these A-T and G-C units *complementary base pairs,* they envisioned them spanning the distance between the two strands like rungs on a rope ladder. Finally, proposing that the paired strands of DNA wind around one another in a spiral arrangement, they christened this two-stranded, twisted model the double helix. The beauty of this model is that it suggests both how the molecule copies itself and how it carries all that vital molecular information.

First of all, the carrier of hereditary information must copy itself accurately, because new copies of DNA are needed whenever a cell divides. The fertilized egg that made you, for example, had to duplicate its DNA to provide identical copies of its genes to each of the trillions of cells that comprise your body. Whenever your body tissues grow to repair wounds or keep up with wear and tear, cells that divide to produce new cells must also produce new copies of their genes. Similarly, when your sexual organs manufacture eggs or sperm, additional copies of DNA are made to pass along to your children. One reason Watson and Crick earned a Nobel Prize was that the structure they proposed neatly explained how those copies are made.

Because the model tells us that DNA's bases always pair up with one another in exactly the same way, we know that if one strand contains the sequence ACGTCTCTATA, its partner must carry the complementary sequence TGCAGAGATAT. Thus, each single strand of DNA contains all the information needed to produce a copy, not of itself, but of its *partner.*

Here's how that information is put to use. When new copies of DNA are required, twenty or more types of active molecules called *enzymes* are mobilized. Some of those enzymes coax the paired bases to "let go" of their partners on the other strand, "unzipping" the double helix and exposing its two single strands. Each of these strands then serves as a template or pattern on which a copy of its partner is assembled.

As the helix unzips, another group of enzymes begins using the existing strands as templates, maneuvering nucleotides produced elsewhere in the cell into position one at a time. One enzyme even "proofreads" the work, making certain that each base pair is properly matched. If a mistake is made, the errant nucleotide is removed and the correct one is inserted. When assembly is complete, the result is a pair of identical, double-stranded DNA molecules, each containing one "old" strand and one "new" strand.

How does this structure store information? The linear sequence of A's, G's, T's, and C's creates a four-letter molecular code in which biological instructions are written. While the form and content of those instructions is unique to genes, they can be compared to more familiar methods of handling information. Human languages, for instance, carry information in sentences formed by strings of words, which are in turn composed of letters. Each sentence specifies action, a subject that carries out the action, and an object that receives the action. "Mary, put the garbage in the trash can," we might say, or "John, put your clothes in the washing machine."

In a sense, the functional units of DNA—the sequences of base pairs we call genes—contain similar instructions, written in molecular language and referring to molecular subjects, verbs, and objects. One gene may carry an instruction that says, "Enzyme 1, unzip the double helix." Another gene might specify, "Enzyme 2, attach these nucleotides together." And so on.[6]

There is one catch. Most languages rely on large alphabets; English has twenty-six letters, Hebrew has twenty-two, and Arabic twenty-nine. DNA seems to have only four. How can apparently simple sequences of A's, G's, C's, and T's convey such sophisticated messages? They can do it because DNA base sequences function like the dots and dashes of Morse code. The Morse system, of course, relies on just two elements—dots and dashes—to represent the twenty-six letters of the alphabet. Thus, the emergency signal SOS is transmitted as (…---…). Because this system is based on only two elements, it requires long strings of those elements to carry information. The word *house,* for example, which contains five letters, requires fourteen Morse elements (….---..-… .).

DNA's four-element code also needs long strings of A's, G's, C's, and T's to encode information; scientists estimate that each of our cells contains roughly 6 billion of them. Even a typical small gene consists of at least 3,000 base pairs, and many genes are much longer.

But how does DNA's coded information actually work? Simply stated, the genetic code directs the assembly of *proteins*—three-dimensional molecules that comprise the physical and chemical fabric of life. In a sense, those proteins are DNA's information translated into flesh and blood.

Proteins, like DNA, are large molecules made up of small units linked into chains. In the case of protein, however, those small units are *amino acids* rather than nucleotides. There are more than forty different amino acids that can be joined to one another in various combinations to form a nearly infinite variety of protein chains. That variety, in turn, allows proteins to perform an extraordinary assortment of tasks. Some function like bricks and mortar to construct physical supports: skin, hair, joints, and nearly everything that holds your body together. Others act like molecular robots; built, programmed, and dispatched by genes, they construct organs, carry messages from head to toe, or enable us to lift our arms. Even body parts that aren't comprised mostly of protein, such as the matrix of calcium compounds that gives bones their strength, are put in place by proteins called enzymes, which hasten chemical reactions. When you devour a candy bar, the changes in your body that help you store extra sugar are guided by still other proteins called *hormones*. Therefore, if DNA controls the production of proteins, it indirectly controls nearly everything that goes on in the body.

Returning to the comparison of molecular information to language, we can liken proteins to sentences, because each of them carries a complete and specific piece of biological information. The amino acids which comprise that protein are the molecular equivalents of letters. Those letters, in turn, form groups that act more or less like molecular words or phrases within the sentence.[7]

Now that you know where the information stored in DNA ends up, you can understand how it gets there—in other words, how the genetic code works. Just as groups of dots and dashes in Morse code represent letters of the alphabet, groups of three bases in the genetic code represent particular amino acids, the equivalents of letters in protein molecular sentences. The DNA base sequence CGT, for example, stands for the amino acid alanine, and the base sequence GAC stands for the amino acid leucine. So the sequence of bases CGT, GAC, CGT on DNA carries coded instructions that represent a sequence of three amino acids: alanine-leucine-alanine.[8]

When all the relevant information in a gene is decoded, much longer strings of amino acids form certain words in protein sentences that specify what a particular protein will look like or determine where it will settle in a cell. Other words and phrases written in the amino-acid "alphabet" direct the action the protein will perform. Still others specify the objects on which the action is to be carried out.

Take the case of hemoglobin, a protein with a straightforward job. Hemoglobin stays inside red blood cells, picks up oxygen molecules when it encounters lots of them (usually in the lungs), and drops them off where there aren't many around (in other body tissues). We might represent the message carried by the hemoglobin molecule with the sentence "I will pick up oxygen where it is plentiful, carry it to places in the body where it is lacking, and will stay dissolved inside the red blood cells all the time." But proteins like hemoglobin not only *carry* directions as written sentences do, they also *perform* the actions the sentence describes. In the molecular world, structure—in this case, the sequence of amino acids in hemoglobin—directly determines function—how the molecule behaves. Line up the amino acids in the right order—in other words, spell the instructions properly—and the protein will do what you want it to. If we can combine our two analogies, the amino-acid letters that write the "sentences" simultaneously build molecular machines that do the work the sentences describe.

THERE'S JUST ONE additional complication. As the improbable chemistry of life would have it, DNA doesn't make proteins itself. Instead, it assembles crews of molecular messengers, decoders, and construction workers to do the job. First, the sequence of bases on DNA is transcribed onto a messenger molecule composed of RNA (ribonucleic acid), a close relative of DNA. This *messenger RNA* carries DNA's instructions to a *ribosome,* a sort of protein factory within the cell. There the sequence of bases is "decoded" and translated into the string of amino acids that defines a particular protein. As the coded message in the messenger RNA is read three bases at a time, the protein chain grows amino acid by amino acid. Step by step, the protein chain grows; letter by letter, the molecular sentence is written.

But the magic isn't over. As the protein grows, it folds up into a complicated, three-dimensional structure. The various parts of that structure (the molecular words we've been talking about) determine the work it does. If a particular string of amino acids was assembled according to genetic instructions for making hemoglobin, part of its program joins it with three other chains to form a complex. The rest of the program instructs it to remain inside a red blood cell, picking up and delivering oxygen as it circulates from lungs to body tissues. The same sort of process, with some differences and embellishments here and there, directs the assembly of every protein in the body.

• • •

THESE MOLECULAR DETAILS may seem to have taken us some distance from the concept of unity, but they haven't. Now that we've made the connection between DNA, proteins, and the processes of life, we can begin to understand both the significance of the molecular revolution and the steps that have made its progress possible. Take hemoglobin, for example. That vital oxygen-carrying protein was one of several molecules that biochemists studied in detail early on, exploring both its physical characteristics and its sequence of amino acids. Some laboratories worked on hemoglobin extracted from human blood, while others focused on samples from a variety of animals. When results were gathered, the unexpected emerged, as geneticist Philip Leder of Harvard Medical School recalled in an interview with one of our producers. "Here's an old book, from 1972," Leder chuckled, pulling a large volume, an *Atlas of Protein Sequence and Structure*, from its shelf.[9] "This would have been one of the first compilations of protein sequence and structure. Just look. Here are human hemoglobins, and here are their structures compared to hemoglobin from whales and horses and seals and kangaroos and goats and camels and chickens. I mean, it's a regular zoo here! And the similarities are surprising."

Again, it wasn't just the *existence* of similarity that was surprising, it was the *extent* of the similarity. By the 1970s, biologists had known for decades that basically the same amino acids were used in the proteins of all organisms. But new data revealed that entire sequences of amino acids, complete molecular words and phrases, were the same in proteins from nearly every organism studied.

Russell Doolittle, a molecular biologist at the University of California at San Diego, remembers the scientific community's surprise as similar proteins showed up in vastly different organisms. "As an example," Doolittle suggests, "let's say we took a human protein with about three hundred amino acids, looked at the lineup, and compared it with the same protein from a bacterium. We would find that nearly half of the amino acids would still be in exactly the same place, in exactly the same order. You expect them to be random but no, they are half identical!"

Of course, not all proteins (or DNA sequences) are that similar across all forms of life. Sulfur-eating bacteria contain molecular instructions for surviving at high temperatures and extracting energy from sulfur that our cells lack altogether. Our cells, in contrast, have both general information on being multicellular and specific information on being human that bacteria don't have. Still, when it comes to the basic mechanics of how to go about being alive, both DNA sequences and proteins of humans and bacteria can be as alike as Doolittle describes.

Among the genes that span the eons between Stetter's bacteria and humans, for example, are those which carry instructions for making ribosomes, the cellular factories that convert DNA's instructions into proteins. In hindsight, researchers might have suspected that ribosomes would look pretty much like ribosomes, regardless of whether they came from bacteria or humans. Protein-making, after all, is fundamental to every living being, so ribosomes must have been "invented" soon after the origin of life as we know it.

Still, Stetter, Doolittle, and their colleagues marvel at how little ribosomes and other cellular components have been tinkered with through all those eons. "It's as if there was one tool kit set down when this all began," Doolittle marvels, "and what evolution has been doing is using bits and pieces from this tool kit to build different creatures. It is just phenomenal. That is why we all love this stuff so much. Speak to any of us, and you will get this same kind of almost awe with respect to the unity of all these molecular systems, from a bacterium to a yeast to a plant to a mouse to a fruit fly to a human." That "tool kit unity," as we might describe it, has been both a major discovery of molecular biology and the source of the field's power.

Why? Because it is one thing to know in principle that DNA carries the information that makes life tick. It is something else entirely to be able to read that information, and something else again to understand molecular sentences well enough to manipulate them usefully. Researchers are already experts at translating the genetic code into the molecular alphabet; every decent high school biology text contains a table that can convert a string of bases into a sequence of amino acids. But in terms of understanding either the molecular words spelled out by that alphabet or the syntax of molecular sentences those words form, scientists are progressing through the "See Dick and Jane run" stage of linguistic development. Biologists still treat life's molecular texts in much the same way that young children treat adult languages: they recognize important words, make reasonable guesses about what those words mean, and try stringing them together into sentences to see if they make sense. Sometimes the resulting sentences work, sometimes they are nonsense, and at other times they are amusing but informative.

There's another appropriate analogy here, one that emphasizes why the unity of life has been so important to molecular biology. You can think of DNA's coded instructions as molecular "software" that runs the "hardware" of life. Just as a word-processing program tells computer hardware what to do, DNA's instructions control life's machinery. Why is that comparison useful? Because if you work with computers the way most of us do, you know enough about your favorite word processor or spreadsheet to *use* it, but you could hardly *write* the program yourself. In much the same manner, molecular biologists know enough about certain DNA-based "programs" to use them without fully understanding how they work.

So it's handy that much of life's software—regardless of the organism it comes from—will run on the hardware of nearly any other living cell. That's why, for example, researchers who discover and learn to control the molecular word processor used by one organism can harness that tool to manipulate genetic text in different organisms without having to learn precisely why or how that particular molecular program works as it does. In fact, virtually all the techniques that comprise the basic tool kit of modern molecular biology were developed by investigators working with viruses and bacteria.

To the uninitiated, neither viruses nor bacteria may seem likely sources of tools with the power to transform the world. Viruses, which are little more than a handful of genes wrapped in a protein coat, lack the molecular machinery needed to perform even life's most basic tasks on their own. Bacteria are more complicated than viruses, but still much simpler than human cells. But both viral and bacterial genes carry instructions written in the universal genetic language.

Viruses use the cross-species hardware compatibility of that common language to infect cells ranging from bacteria to humans, and to take over the molecular machinery of their hosts, forcing it to manufacture more viruses. For their part, bacteria, which have probably been plagued by viruses for longer than most other organisms, have evolved genetic instructions to help them resist those hostile takeovers. Intriguingly, bits and pieces of those genetic instructions—molecular programs for controlling genes that have been evolved and perfected by nature herself—have served, both directly and indirectly, as genetic engineers' principal tool boxes.

Back in the 1960s, researchers observed that DNA didn't last long when mixed into a "soup" made from homogenized bacteria. They also noticed that almost any kind of foreign DNA taken into living bacteria rapidly broke into pieces. As it turned out, these experiments had uncovered one of bacteria's molecular defenses against viral attack. This weapon, an enzyme, managed to "chop up" the DNA of invading viruses and render it harmless without destroying the bacteria's own DNA. It accomplished this feat by cutting the double helix only when it recognized a specific sequence of bases that was found in viral DNA but not in bacterial DNA.

Since that first discovery, no fewer than eighty different enzymes of this type have been found in various species of bacteria. Each one recognizes a particular series of bases on DNA—the sequence GAATTC, for example—and cuts the double helix only where that sequence appears. Because these enzymes disable or "restrict" DNA, they were called *restriction enzymes*.

Once word of restriction enzymes circulated, molecular biologists began dancing around their laboratory benches. Why all the excitement? Because biologists had known for some time that in order to study and control DNA, they had to be able to manipulate it, cut it into pieces without destroying the information it carries, alter it, and put it back together. But they knew so little about molecular machines that they hadn't a clue how to build and program a simple set of molecular "scissors" and "glue" to cut and paste DNA purposefully. (Investigators had long since come up with myriad ways to hack the double helix to pieces indiscriminately, but that's much less useful.)

The discovery of restriction enzymes solved that problem. As luck would have it, when restriction enzymes cut DNA, they leave "sticky ends" at the site of the cut. The ends are called sticky because one piece of DNA cut by a particular restriction enzyme tries to pair up again with another piece that has been cut the same way. Thus, with several specific restriction enzymes in hand, researchers can say, "I want to cut out this piece of DNA *here* and stick it in *this* position, and I want to attach this other piece over *there*." Few, if any, of the applications that fill this book would be possible without this ability.

Another vital tool was commandeered directly from a peculiar group of viruses that carry their genetic information in the form of RNA rather than DNA. Back in the 1950s and 1960s, the "central dogma" of molecular biology held that information flow in living systems always proceeded in one direction: from DNA to RNA to protein. But around 1970, two independent research teams working with a small and obscure group of RNA viruses realized that those parasites seemed to ignore that rule; they somehow made DNA copies of their RNA genes. Incredulous scientists called these viruses *retroviruses* because their activity was seen as retrograde, or backward.

Retroviruses, which cause diseases ranging from certain cancers to AIDS, led molecular biologists to *reverse transcriptase,* an enzyme that does what its name implies; working on a strand of RNA, it assembles a complementary strand of DNA. At the behest of viruses, this enzyme produces viral genes on strands of DNA that insinuate themselves into host chromosomes, causing essentially incurable infections. In researchers' hands, reverse transcriptase enables new manipulations of genes that make both genetic engineering and gene therapy possible.

Many other tools were discovered in similar ways, but one in particular has proved invaluable. Ever since researchers first began tinkering with DNA, they have needed to copy selected pieces of the double helix in order to obtain enough material to study. Early on, they relied on bacteria to do the job for them. After isolating a gene, they would add it to the genome of a bacterium, encourage that bacterium to multiply, then extract many copies of the sequence desired. That technique—the source of the original phrase "cloning a gene" because it involves clones of bacteria—works because bacterial machinery copies a human gene, or any other gene, just as it copies its own. Although genes are still sometimes cloned in bacteria, there are some situations in which researchers want easier or more precise ways of copying DNA.

Other techniques now exist because investigators eventually realized that they didn't need the entire bacterial cell to do the job if they could isolate the enzymes that do the work. Those enzymes, called *DNA-polymerases,* work on any single-stranded DNA they encounter, using it as a template to assemble a complementary strand. The basic technique employing polymerases was simple. A single-stranded copy of a gene was mixed with polymerase and a supply of individual nucleotides. Under the right conditions, the polymerase did its work, copying each strand and producing a double helix. If researchers could find a way to repeat this process many times, they could double their DNA with each round, ultimately producing thousands, or even billions, of copies.

There was only one catch. Polymerase works only on single-stranded templates; it can't unzip the double helix by itself. Investigators knew that gentle heating could unzip the newly assembled helix, exposing single strands to polymerase once again. But the amount of heat needed to do that also permanently destroyed the activity of ordinary polymerase. As a result, each time they wanted to repeat the process, they had to add a new batch of enzyme, a cumbersome and expensive requirement.

The solution to this predicament was found in the cellular machinery of hot spring bacteria. Once researchers learned that these organisms thrived at temperatures that would literally cook most other cells, they realized that those bacteria must have DNA-duplicating enzymes that worked under such conditions. Sure enough, when hot spring polymerase was isolated and tested, it retained its activity at temperatures that unzip normal DNA. Even though the enzyme came from organisms not unlike Stetter's bugs, it recognized DNA from other organisms and performed its appointed task. Today the DNA copying procedure known as *PCR,* for Polymerase Chain Reaction, is straightforward and easily automated; the ingredients are mixed together, then heated and cooled cyclically until they produce as many copies of genes as needed.

As these and other molecular tools fell into place throughout the 1960s and 1970s, researchers gradually learned how to use them in various combinations to isolate genes, move them around, read their base sequences, and study their proteins. By the 1980s, techniques that were once subjects of doctoral theses were standardized and collected into do-it-yourself "kits," many of which can be used by well-trained technicians with a high school education.

Enormous amounts of data began to accumulate, far more than could be stored and accessed efficiently by using old-fashioned filing techniques. Fortuitously, the maturing field of computer science offered new ways to handle them. Today, computer databases of DNA and protein sequences allow researchers to compare molecular information more quickly and easily than ever. Not surprisingly, they now make those comparisons as a matter of course.

"It's second nature," Phil Leder notes. "You clone a gene, a sequence that you think is interesting. You predict the amino acid structure of its protein and rush off to the computer, which can tell you, in something like two or three minutes, whether what you've got is like anything else that has ever been sequenced before. And often you discover that it has already been cloned by somebody else in a weasel or something like that. Then you find that the gene is present in every single vertebrate species from dolphin to human being. It's absent in alfalfa, by the way. But yes, the computer tells you, 'Your sequence is known in such and such an enzyme.' Then you already know something about what you are dealing with. That's a lot of information right there. That's tremendous power."

OF COURSE, many genes are still new to science, so they aren't yet found in databases. But the molecular unity of life enables researchers interested in complex problems to look for answers in organisms that are simpler to study and easier to work with than humans, monkeys, or even mice. Why should they need to go to simpler cells? Because, according to Barry Polisky of Indiana University, biologists' lack of knowledge about molecular machines puts them in roughly the same position as a crew of budding auto mechanics who, without the aid of repair manuals, try to learn how cars work by fiddling with them. There are plenty of models to choose from, all of which share a multitude of features. "You have to ask yourself," Polisky notes, "what kind of car do you want to start with? Do you want to work on a Ferrari with 100,000 little things attached to the engine? Or do you want to work on a 1948 Ford truck, where you open the hood and there is nothing there but the engine? That's the way we chose what to work on."

By way of example, Gerry Fink relates a story of researchers studying an inherited cancer, neurofibromatosis, which, like all cancers, is a very complex disease. But the root of all cancers is cells that grow out of control owing to one or another faulty molecular component that keeps cell division machinery operating in high gear even when it shouldn't. The neurofibromatosis team thought they had isolated a gene that had something to do with the operation of cell division. They felt certain that this gene causes neurofibromatosis in humans, and they managed to determine its base sequence. But exactly what did that gene do? How did its protein cause the disease? When they took their information to the database, the result was inconclusive; the computer said "not sure."

At that point, they decided to put their gene into yeast to see what happened. In a way, that was like attaching a tire or steering wheel from a Ferrari to a 1948 Ford truck, or even a go-cart, to figure out how the part works. But the molecular machinery of cell division in yeast cells is sufficiently similar to ours that the results were informative. "There were several mutants in yeast," Fink explains, "where cell division is abnormal. And it's a very simple thing to see. So they put the human neurofibromatosis gene into the mutant yeast. And it worked! The abnormal yeast cells changed their function in such a way that [the researchers] could figure out what their gene was doing. So those guys zoomed ahead in their work. They could have spent *years* trying to figure out function by doing biochemistry—asking what reactions does the protein catalyze, and so on—but they got past all that by putting a human gene into yeast!"

• • •

DISCOVERIES LIKE THE neurofibromatosis gene are both common and visible these days. At least once a week, headlines trumpet accounts of researchers who have tracked down yet one more gene that either leads directly to a specific disease or predisposes its bearers toward illness. Words like "discovery" and "tracking down" are appropriately used in this context, for gene hunters are very much like explorers prospecting in uncharted territory. If anything, gene hunters' work is harder, because it requires two parallel types of inquiry.

First, like any explorers, gene hunters survey as they go, compiling "maps" of chromosomes on which they record easily identifiable landmarks and the locations of specific genes. That work is important, because in many ways, finding a gene in the human genome is like trying to track down an individual person with no information other than the fact that the target is located somewhere on earth. At first, researchers hunting for genes had only one choice: to chop all our DNA into tiny, gene-size bits and randomly test each piece to see if it was the right one. Given the size of the human genome (3 billion base pairs, remember), that's roughly akin to mounting a manhunt by wandering randomly through the world and reading the names on all the mailboxes.

Second, after locating genes, researchers want to "read" them, and figure out how all the information they contain makes life work. Again, the size of the task is daunting; 3 billion letters of average size, printed on standard typing paper, would fill roughly 2 million pages. That's a lot of reading material. To confuse matters further, a surprisingly large part of our DNA carries base pairs that don't seem to be part of functioning genes—that's reams and reams of sequences that are either gibberish or peculiarly constructed molecular sentences whose meaning we don't yet understand at all. By current estimates, scattered among all those apparently meaningless pages—called junk DNA by some researchers—are somewhere around 100,000 different human genes. So here we are, at the equivalent of a second-grade reading level when it comes to molecular language, trying to make sense of a text whose size makes the collected works of Shakespeare look like a short story.

However, length isn't the only issue. To understand the full extent of the challenge, imagine that we are extraterrestrials who've discovered Shakespeare's plays written in code, inscribed on pages that have been ripped from their binding and shuffled, and peppered with long stretches of nonsense syllables. Suppose we suspect that the outcome of one particular play is tragic. Wondering why that is so, we would like to rewrite it to give it a happy ending. How do we proceed?

The first step is to decode the text and rewrite it as a stream of letters. But because the pages are scrambled, we have to dive in and choose strings of code more or less at random. Assume that the first sentence we decode is "oromeoromeowhereforeartthouromeo." We break that string into words and translate it into our own language. Our reward is "O Romeo, Romeo! wherefore art thou Romeo?" The translation presents scores of questions. Who is speaking? Who, or what, is Romeo? What is the relationship between the groups to which the speaker and Romeo belong? Would the story still proceed without this sentence? As we try to solve the puzzles, we decode sentence after sentence, probing for common words and themes. We string phrases and sentences together, cut them apart, and rearrange them to see how they influence one another. The word *Tybalt* that we often find in other sentences seems similar in some ways to the word *Romeo;* would it change the plot if we switched them around? What can we learn from this story that might help in understanding others? We constantly search for "stage directions" in the script, as scrambled and as complex as it may be.

Until the late 1980s these formidable tasks—finding, mapping, and reading genes—were done as much of biological science is done—independently, in laboratories working on a host of unrelated projects. Collaborations were individual events; gene hunters who had already searched some part of a chromosome would share data with others, helping them with the technical equivalent of "I've been down that road, and your man isn't there." Computer databases made that sort of exchange easier and more frequent, but bits and pieces of the map were still plotted fairly haphazardly.

So in the interest of efficiency, many prominent molecular biologists have rallied around one centralized project that some refer to as its Holy Grail: a mission to draw accurate maps of all human chromosomes and sequence the 3 billion bases that constitute our genetic endowment. That effort, united under the Human Genome Project, is based on the recognition that genetic similarities among us all ensure that we do share a common human genome. Yet its principal goal is to find, within that similarity, the minute differences in DNA sequences that predispose individuals to a wide range of inherited and degenerative diseases.

The project aims first to create a *genetic map* that will serve as a molecular road atlas to help researchers find particular "addresses" in the genome more easily. That map will be based on several thousand specific sets of base sequences, or *genetic markers,* strung out more or less evenly along the twenty-three pairs of chromosomes. Finding those markers isn't technically difficult these days; but it is tedious and time-consuming. It requires finding very large families—researchers in Utah are making good use of extended Mormon families, for example—analyzing their DNA, and looking for genetic landmarks located in various parts of the genome. Once enough of those landmarks are identified, they will help direct researchers to the genetic equivalents of the right "country," then to the right "city," and ultimately to the neighborhood of particular genes.

At the same time, other teams are working to produce what they refer to as a physical map of the genome. This work requires chopping chromosomes into large, overlapping pieces, cloning those pieces in bacteria or yeasts, then determining how those pieces fit together. By doing so, researchers divide an enormous length of DNA into manageable pieces that can be located and copied easily for more detailed study. Ultimately, the existence of both maps will enable a researcher in search of a particular gene to locate it through the genetic map, then make copies of it to study with the aid of the physical map.

According to the genome project's original proponents, one main advantage of doing this important but tedious work in an organized and concerted fashion, rather than in bits and pieces, is to create an environment that encourages the development of streamlined, automated techniques to speed the process and lower its ultimate cost, originally budgeted at $3 billion. At first blush, that strategy is working. Many who criticized the project at its inception in 1990 originally considered its goal to map the entire genome by 1995 to be overly ambitious. Yet by the end of 1992, researchers finished high-resolution physical maps of two entire human chromosomes, the Y chromosome and chromosome 21, an achievement signaling that the larger mapping goal is well within reach. The task of sequencing all, or at least the non-junk parts, of the genome, on the other hand, is further off—2005 by the project's original reckoning—but similar advances in sequencing technology are speeding up that work as well.

Most genome project advocates see its major goal as the production of biomedical crystal balls. Because genes determine how our bodies work, finding and understanding their faults could enable physicians to predict disease and intervene on patients' behalf. And because many common and serious ailments, such as cancer, arthritis, cardiovascular disease, and multiple sclerosis, are caused by several interacting genes, the more complete our knowledge of the human genome, the better our chances to fight them.

The project's strongest proponents, people like Leroy Hood of the University of Washington in Seattle, are enthusiastic believers in the positive power of genetic medicine who are deeply committed to this humanitarian vision. "My God," Hood almost shouts in his enthusiasm, "once we've identified the genes that control these traits, and understand their function, we'll be able to repair their defects. And what we'll do is when you're born, we'll take your DNA and look at a hundred disease-predisposing genes, and we'll identify your pattern of predisposition for disease, and then institute appropriate regimens so these things will never come up. Once we're able to do that, we've revolutionized medicine by these preventive techniques. I see this as the real key to gaining control of the escalating costs. That's the future, and it's very exciting."

Technologies that go a long way toward making Hood's dream real are already in the wings, in the form of techniques to make identifying genes easy, fast, and reliable. One biotech company has the technology to miniaturize hundreds of DNA tests on a single silicon chip. That leads Hood to predict a diagnostic dream come true. "I could envision," he explains, "a chip where you have the assays you need for two hundred common infectious diseases, the assays you need for the fifty most common cancer-causing genes, the assays you need for forty common genetic diseases, and the assays you need for one hundred disease-susceptibility genes. These would just be little chips that could be routinely tested against a patient's DNA, and you would get these very clear, plus/minus answers as to what they had."

While "the" genome project chips away at the human genome, of course, other researchers focus on other organisms. The "worm project," for example, concentrates on the DNA of a tiny, transparent worm called a nematode, a favorite experimental subject of molecular biologists and an animal with a more modest genetic endowment. By way of comparison, the 3 billion bases that we've said make up the human genome, typed out on 2 million pages, would take up about 415 linear feet of filing space if you could cram 200 pages into every inch. The nematode genome, on the other hand, which consists of "only" 100 million bases, would occupy little more than 14 feet of space. The genes of a typical bacterium, on the same scale, could be "filed away" in only about 8 inches of space. And a virus, whose genome would be no more than a very few pages, would fit in a single manila folder. Once again, since the basic instructions of life are so similar, a great deal of information gleaned from these more manageable molecular texts will be directly applicable to human affairs.

While the genome project is driven—and funded—primarily by researchers and institutions aiming to predict patients' medical futures, some scientists envision the information it gathers as a treasure trove for studying the past. The University of Munich's Svante Pääbo calls the genome project "the greatest archaeological excavation of all time" because of the light it is likely to shed on the history and evolution of our species. The main project, with samples drawn from European and American Caucasians, sports side branches, which aim to collect and "bank" blood samples from aboriginal peoples around the world. Comparisons of similarities and differences among ethnic groups promise to facilitate studies of historical relationships among modern human populations. And significantly, broader comparisons between DNA sequences of humans and those of other organisms have already begun to expand our understanding of the history of all life on earth.

Through these comparisons, many molecular biologists have been drawn—some willingly, others kicking and screaming—into the study of life's past. Until the mid-1970s, their studies of molecular systems concentrated on very clear, very specific questions. They wanted to identify the carrier of genetic information and to understand how that information is read. They were totally unconcerned with historical events. But along came the sorts of similarities among life's parts and tools that surprised

Stetter, Doolittle, and Fink, and which ultimately forced researchers to ask, "Why is so much of the living world organized in this particular way, when it could just as easily, or much more easily, have been some other way?"

The answer to that question was the fact of evolution. Overwhelming similarity at the molecular level testifies that life's information storage system appeared only once. Our genes resemble those of other organisms because the genes in all of us are descendants of the genes once carried by our common ancestor—life's common denominator—who lived more than 3.5 billion years ago. The line of descent is unbroken, the web of relationships irrefutable. Ways of performing the tasks necessary for life were invented over evolutionary time, and the instructions for all of them were recorded and preserved in the immortal helix that unites us all.

The implication of that shared history is staggering. Until now, biologists have traced the history of life by comparing skulls and fingers, fur and feathers, teeth and claws, of living and fossil organisms. But hidden in the DNA of yeast and fly, mouse and human, is information of incredible antiquity and enormous value, information whose richness and texture far surpass the observable physical features of organisms.

"There's the hope," explains molecular evolutionary biologist Rob Dorit of Yale University, "that recorded at the molecular level is the deepest history of organisms. The contention that long after what I will call surface similarities have vanished, there is still the wake of history etched in the deepest comparisons that you can possibly imagine." Information extracted from the coils of the double helix has already begun to revolutionize the way we reconstruct life's family tree. And, as you will see in the next chapter, it has also produced both a dramatic reification of Charles Darwin's genius and a profound restructuring and extension of classical evolutionary theory. Ultimately, it seeks to explain how sulfur-eating bacteria and the peculiar species *Homo sapiens* both came to exist—and continue to persist—on earth.

2

Vive la Différence

Variety's the very spice of life.

William Cowper

AT THE BEGINNING of the nineteenth century, Scottish adventurer Mungo Park landed near the mouth of the Gambia River in what is now Senegal. Funded by the Royal African Association to explore West Africa's interior, he planned to march upriver along the Gambia, cut across to the upper reaches of the Niger River, and sail downstream to the Gulf of Guinea. Park equipped himself well, with forty-five men, plenty of pack animals, and ample provisions. But he made one fatal error: he began his journey at the beginning of the rainy season, when mosquitoes swarmed everywhere.

The expedition ran into trouble almost as soon as it set out, not primarily from hostile local people, but from a disease called the fever. According to historian Philip Curtin of Johns Hopkins, the daily journal Park kept for his sponsors was filled with matter-of-fact accounts that showed how quickly death became part of the daily routine. "One day's entry," Curtin recalls, "would read something like, 'We made camp last night. Three men died.' Two days later, he might have entered, 'Today we bought a donkey, camped by a river, and five men died.' Similar disasters would happen step by step along the trip. By the time they reached the Niger, two and a half months later, only five men were left—and they were already sick."

Park's ill-fated comrades weren't alone; the vast majority of Europeans who traveled through West Africa in those days died from the fever, or malaria, as it is known today. The region's reputation as "the white man's grave" acknowledged what was then a mystery and is still a fact of life. While European explorers perished in droves, ebony-skinned

Africans survived handily, striding alongside white men so weak that they could barely cling to their horses. It wasn't that local people never contracted malaria; some certainly did. Yet many of those who were infected didn't become as ill as Europeans, and far fewer of them died from the disease. What was behind this mysterious, apparently innate protection against malaria?

In 1904, roughly a hundred years after Park's expedition and many miles away, Dr. James Herrick was confronted by a far different medical problem. He described a patient with striking symptoms. His legs and thighs were covered with scars as large as silver dollars; his heart was enlarged and beat with a murmur; he had fever and chills. At first Herrick could find no cause for the man's complaints. But when he made a routine smear of the young fellow's blood, he saw something he had never before seen. "The shape of the red cells was very irregular," he reported to the *Archives of Internal Medicine,* "but what especially attracted attention was the large number of thin, elongated, sickle-shaped and crescent-shaped forms." Although Herrick had no way to prove his suspicions, he concluded his report with the prescient suggestion that "some unrecognized change in the composition of the corpuscle itself may be the determining factor."[1] Despite this insight, Herrick could do nothing for his patient; the man died of immune system collapse tied to a damaged spleen.

Over the next few decades, other physicians noticed that this syndrome, which came to be called sickle-cell anemia, runs in families and seems to be most prevalent in people of African descent. Researchers confirmed Herrick's suspicion that sickle-cell patients' hemoglobin, the oxygen-carrying protein in red blood cells that we discussed in Chapter 1, seemed to differ from ordinary hemoglobin, though they couldn't say precisely how or why. What was the cause of this debilitating and often lethal condition? And how did it come to be particularly common in people belonging to certain ethnic groups?

Years later, the last pieces of these two apparently separate puzzles fell into place, revealing that a single phenomenon explains them both. The ability of West Africans to resist a lethal infectious disease, malaria, and their prospect of inheriting an equally lethal ailment, sickle-cell anemia, turned out to be flip sides of the same molecular coin. The full, fascinating story combines discoveries in molecular biology with efforts to explain evolutionary change that reach back to Charles Darwin himself. In the process, it provides a window onto studies of a central question in the biological sciences: How, although like begets like, has life changed over time and produced the extraordinary diversity of organisms that inhabit our planet today?

• • •

LONG BEFORE GENES were understood, Darwin recognized that inheritable variation was both essential to and an integral feature of life as we know it. This apparently simple observation represented a major and controversial philosophical leap for Darwin's time. Previously, most Christian theologians had argued that because God was perfect, all His work had to be perfect. They further argued that this perfection implied stability, for perfect creations should not change. Devout scientists, in turn, asserted that variations in plants and animals, representing imperfections and deviations from God's eternal model for each species, existed only because the material world was corrupt.

Darwin turned that view on its head. Beginning *On the Origin of Species* with a chapter entitled "Variation under Domestication," he noted that traits of cultivated plants and animals vary all the time. Even among lambs born to a single ewe, some might have wool that was lighter or darker, longer or shorter, thicker or thinner, than that of their siblings. Farmers, recognizing that much of this variation was passed from parent to offspring, chose the most desirable variants for breeding. By so doing, Darwin explained, breeders improved their stock. He wrote, "The key is man's power of accumulative selection: nature gives him successive variations; man adds them up in certain directions useful to him."[2]

Having described how artificial selection produces new varieties in captivity, Darwin sought an analogous process to explain change and the appearance of new species in nature. His challenge was to find a defensible scientific mechanism to account for both the diversity of life and organisms' remarkable "fitness" or adaptation to their particular environments. Ultimately, Darwin settled on a phenomenon he called *natural selection,* which he explained as follows:

> As many more individuals of each species are born than can possibly survive, and as, consequently, there is a frequently recurring struggle for existence, it follows that any being, if it vary however slightly in any manner profitable to itself under the complex and sometimes varying conditions of life, will have a better chance of surviving, and thus be naturally selected.

> From the strong principle of inheritance, any selected variety will tend to propagate its new and modified form.[3]

This was, in fact, Darwin's unique contribution to biological thought. He was not the first, by a long shot, to suggest that life had changed over time. But he was the first to marshal decades of observation and analysis to propose and support a plausible scientific mechanism by which evolution could occur. Biologists have spent more than a century studying, revising, and expanding upon Darwin's work. For most of that time, the *fact* of evolutionary change—the reality that organisms have changed during earth's long history—has been universally accepted by scientists. But knowing that evolution *has* occurred is very different from understanding *how* it proceeds.

Today, thanks to studies in genetics, biochemistry, and molecular biology, we can see into the molecular events behind evolutionary change in a way that Darwin never could. We know that heredity is based on DNA; wildebeests give birth to wildebeests and zebras sire zebras because parents pass genes along to their offspring. We know that the heritable variation Darwin described but could not explain is based on variations in DNA; among those wildebeests and zebras, no two are exactly alike because each of them carries a unique combination of genes. We also know, as Darwin did, that organisms do not and cannot evolve because they "need" or "want" to, and that natural selection cannot create mutants to serve its purposes. Darwin knew little about heredity, and nothing about genes, but he recognized that inheritable variation is random. Today we know that genetic variants either arise or do not arise strictly by chance.

The source of all that variation is twofold. Much of it comes from sex, or more precisely, from sexual reproduction, during which a great deal of gene shuffling goes on. As Gregor Mendel guessed, typical plants and animals contain two sets of genes, each contributed by one parent. In our species, that genetic endowment is borne on twenty-three pairs of chromosomes. When testes or ovaries make reproductive cells, one member from each of those pairs is shuffled randomly into sperm or eggs with one member of every other pair. That process alone means that any human parent can produce an astronomical 2^{23}, or 8.4 million, different combinations of genes in their reproductive cells. Yet sex actually creates many more gene combinations than that, because gonads don't pass along parental chromosomes to offspring intact. Instead, before eggs and sperm are formed, the paired chromosomes that will ultimately be divided between them line up side by side and swap pieces with their partners at random. Such gene swapping between chromosomes is one reason that genetic mapping bedevils the researchers you'll meet in later chapters. It also helps explain why no two siblings, except for identical twins, ever look exactly alike; shuffling genes into different associations means that you can end up with your mother's eyes, your father's nose, and hair that combines traits from both.

Still, life wouldn't change—our primate ancestors wouldn't have evolved into modern humans—if groups of identical genes were simply shuffled around indefinitely. There's another source of inheritable variation, one that alters the information genes carry. As meticulous as DNA replication is, mistakes in the process occasionally occur. Sometimes a single base is changed, from an A to a T perhaps. Other times a base may be dropped from the sequence. Still other times strings of bases are stuck in the wrong places, inverted, repeated, or left out. Although "proofreading" enzymes usually correct such blunders, every now and then an error slips through. We call such mistakes *mutations* (from the Latin word meaning "to change"), and refer to individuals who carry them as *mutants*.

Much of this book deals with mutations in the human genome that either cause inherited diseases such as cystic fibrosis or lead to the ravages of cancer or multiple sclerosis. But mutations aren't always bad, particularly from the standpoint of species survival over evolutionary time. Some mutations—especially, but not exclusively those which occur in "junk DNA"—may have no noticeable consequences. More important, although other mutations may cause individual organisms to suffer or die, they generate a critical element of inheritable variation. That variation, often referred to as genetic diversity, provides the "raw material" upon which natural selection operates. That's important, because earth's environments are constantly changing, often unpredictably. If random mistakes in DNA didn't enable species to adapt to those changes, organisms couldn't evolve—and might not survive.

How could a minor change in a gene enable an organism—or more accurately, a population of organisms—to survive an otherwise lethal change in its environment? Interestingly, the case of malaria and sickle-cell anemia offers an example of precisely that situation, for it centers on just such a mutation. The story begins with a change in DNA that is as minute as you can imagine: from among the 6 billion base pairs present in every human cell, from among the hundreds of base pairs that comprise the gene that directs the assembly of each hemoglobin chain, one single, solitary base pair is different. That one alteration in base sequence alters the message carried by messenger RNA, causing ribosomes to attach a different amino acid—valine instead of glutamine—into the sixth position of the protein chain. The result is that instead of producing ordinary hemoglobin, this set of molecular instructions produces a slightly different molecule, *Hemoglobin-S*. Because this altered stretch of DNA represents a specific version of the hemoglobin gene, it is often called the Hemoglobin-S allele. Those who carry it also carry Hemoglobin-S in their red blood cells.

What difference could a single amino acid make in a chain consisting of hundreds? It turns out that the rules of molecular "grammar" are far less forgiving than those of human languages, so the slightest mistake in spelling or punctuation can cause a profound change in the meaning of a molecular "sentence." In physical terms, even the smallest alteration in amino acid sequence can dramatically alter the function of the entire protein chain.

There is no way to demonstrate with an English sentence the true nature of the change wrought by that single-base mutation in the hemoglobin gene, but we'll get as close as we can. We suggested in Chapter 1 that the protein sentence carried by ordinary hemoglobin might read, "I will pick up oxygen where it is plentiful, carry it to places in the body where it is lacking, and will stay dissolved inside the red blood cells all the time." The Hemoglobin-S sentence, on the other hand, might say, "I will pick up oxygen where it is plentiful, carry it to places in the body where it is lacking, and will *not* stay dissolved inside the red blood cells all the time."

In fact, that's close to what actually happens. Hemoglobin-S molecules are slightly less soluble than ordinary hemoglobin, so they come out of solution under certain conditions. When that occurs, the molecules bind together in long chains and stretch normally flexible red blood cells into rigid sickle shapes like those Herrick saw in his patient's blood smear.

People who, like Herrick's patient, suffer from sickle-cell anemia carry two copies of the Hemoglobin-S allele, which means that they received one copy from each parent. Both their hemoglobin genes carry instructions for making Hemoglobin-S, so their red blood cells are filled almost entirely with that altered protein. When the amount of oxygen around their red blood cells drops below a certain point, as it does during strenuous exercise, Hemoglobin-S comes out of solution, forcing the cells to sickle. Then, as those sickled cells enter the body's smallest blood vessels, they get stuck, block the flow of blood, and cause tissue damage.

Individuals who carry only a single copy of the Hemoglobin-S allele received from just one of their parents, however, don't have that serious problem. Why? Because those individuals also carry an unaltered copy of the gene that came from their other parent. When the time comes to make hemoglobin to supply red blood cells, both copies of the gene "turn on." Thus, roughly half the messenger RNA they produce directs the production of ordinary hemoglobin. Because only half the hemoglobin in their red blood cells is Hemoglobin-S, those cells are less likely to sickle.

This molecular explanation begs an important question: How did the Hemoglobin-S allele become common in certain groups of people? The path that ultimately led to our current understanding began in 1946 in what is now Zimbabwe, when an observant physician thought he saw a correlation between Hemoglobin-S and resistance to malaria. According to his observations, children who carried a single copy of the allele seemed more resistant to malaria than children who didn't carry it. These kids could still be infected by malaria, but their symptoms were less severe and their infections were less often fatal.

Eyes wide, researchers scurried to maps of the globe, on which they recorded both the geographic distribution of malaria and the known occurrence of the Hemoglobin-S allele. They discovered that nearly everywhere malaria was common, the allele was present in a significant proportion of the population. In addition to people of pure African descent, it was found in people of varying skin shades in Italy, Greece, and Portugal and across a variety of ethnic groups in the Middle East and Asia—in short, wherever malaria poses, or has posed, a persistent threat to life. Conversely, where the disease was absent, the allele was also missing. The only exception to this relationship is the allele's presence among African-Americans transported en masse from malarial homelands to the New World by the slave trade.

Given this background, medical historians and evolutionary theorists propose the following scenario for the origin and spread of the sickle-cell trait. At first, the virulent form of malaria we fear today did not infect humans. During that time, the Hemoglobin-S mutation occurred here and there in various human populations around the world. (Studies of living populations suggest that it cropped up three times in Africa alone.)

But before the deadly form of malaria affected people, the allele, although it caused no harm to those who carried it, conferred no benefit either, and probably remained relatively rare.

Then about 10,000 years ago, *Plasmodium falciparum*, a form of malaria that had previously infected only birds, made the jump to humans. It is difficult for those of us living in temperate-zone countries to realize how devastating that event was. Since its appearance, falciparum malaria has been one of the greatest scourges of human populations. Known in ancient Assyria, Egypt, and China, it attacked humans wherever they lived in large groups. Once the slave trade carried malaria to the New World, it played a major role in annihilating the indigenous people of Mexico and Central and South America. Even today, malaria kills 1.2 million people annually, more than any other single disease.

In Africa and many parts of Asia, the disease is still a very real problem, as patients in the Albert Royer Children's Hospital in Dakar can attest. One rail-thin child, nearly unable to stand and walk without his father's help, looks confused and disoriented as two physicians, Mohamadou Fall and Dembel Sow, examine him. Fall speaks softly to the boy and his father in French, reassuring them as he takes a case history. A third physician, Ronald Nagel of Albert Einstein College of Medicine, explains the boy's serious case. "This is cerebral malaria, the most deadly complication of malaria," he says quietly, shaking his head. "It happens when red cells with the parasite inside clog up the small vessels in the brain. Suddenly the brain doesn't have enough oxygen. That creates a very severe problem, and if it is not diagnosed early and if it is not treated early, it can kill the patient. This is by far the most common way in which malaria patients die."

Moving through the unit, apparently accustomed to the cries of babies and young children that fill the air, Nagel pauses by another bedside. "Fortunately," he remarks, smiling at the weak but relaxed young girl in the center of the bed, "in this case the diagnosis was early and the treatment was very effective and rapid. She has a very good chance of surviving without any defects. But of course, only time will tell."

The severity of these children's afflictions raises the question of why malaria is so deadly. The problem lies in the life cycle of the parasites that cause it, single-celled members of the genus *Plasmodium*. Once transmitted to a human by the bite of the appropriate mosquito, those parasites spend a short period in the liver before emerging into the bloodstream and searching for red blood cells. The parasites attach to their cellular targets by latching on to particular proteins on the cell surface. They then enter the cells, feed on hemoglobin, and multiply rapidly. Soon they break free, destroying host cells in large numbers and causing bouts of raging fever alternating with chills. The newly released parasites seek out new red cells to invade within hours, and the cycle continues.

Because *Plasmodium* spends nearly its entire life *inside* our cells, it is effectively hidden from our normal biological defenses against disease. Our immune system can "see" the invaders only for brief periods between the time they break out of one set of red blood cells and the moment they enter their next victims. Like an army of guerrilla fighters hiding in an impenetrable jungle, *Plasmodium* keeps the body outfoxed and off balance. That's why, when people who have never been exposed to malaria become infected for the first time, the course of the disease is usually rapid and violent, and often fatal.

But here's where Hemoglobin-S enters the picture. Studies by Nagel and other researchers suggest that Hemoglobin-S mixed with ordinary hemoglobin hampers the malarial parasite in two main ways. First, when the parasite begins to grow inside the red blood cell, it changes the cell's internal environment in ways that cause Hemoglobin-S to sickle. This increases the likelihood that infected red cells will be removed by the body's "quality control check" as they pass through the spleen. As the spleen removes those sickled cells, it kills the parasites before they can mature. Second, *Plasmodium* doesn't seem to be able to eat Hemoglobin-S very well, and a significant number of parasites starve to death, giving the patient a much better chance of surviving.

The net result of these effects confirms the observations of that prescient doctor in Zimbabwe half a century ago; children who carry a single copy of the Hemoglobin-S allele are much more likely to survive their first malaria attack. That's a vital advantage, because once a child passes the age of five, he or she is usually protected from life-threatening infection by the "education" of the immune system we'll describe in Chapter 5. The trick is to survive long enough to acquire the active immunity. According to Nagel, a single copy of the Hemoglobin-S allele gives children a vital chance to build and maintain the immunity that protects them during later life.[4]

Historically, once malaria appeared, it acted as a form of natural selection on human populations. In areas where malaria was rampant, people who just happened to carry the allele for this new form of hemoglobin were more likely to survive and bear children than people who didn't carry it. The blind-luck aspect of this situation is important to keep in mind. There was no way for those individuals, or their DNA, to "know" that any particular genetic variant would have lifesaving potential. "The sickle-cell gene did not arise because there was malaria around," Ronald Nagel emphasizes. "It might have occurred many times over the history of mankind, but it was never useful until malaria came along. Only when that mutation became an advantage to the carrier did the numbers of people carrying it increase, because those people who lived longer could produce children and pass the gene along to their progeny."

As carriers of the Hemoglobin-S allele grew in numbers, the frequency of the protective allele rose rapidly in the population. At first, while the allele was still rare, people who carried one copy invariably had children with partners who did not. By the laws of genetics, half the children such a couple produced carried a single copy of the mutant allele, and they were protected from malaria at little or no cost to their health. That's why in Dakar today, when Nagel, Fall, and Sow admit a patient with malaria, they test for the presence of the Hemoglobin-S allele. If they find it, they are much relieved, for it means that their patient has a more favorable prognosis.

But people in Senegal and other malarial regions must also deal with the negative effects of this mutation. Over time, as more carriers of the Hemoglobin-S allele survived, its frequency in the population climbed to the point that people who carried it began to marry one another. Half the children of such parents receive a copy of the allele from only one parent, and they are protected against malaria. But one-quarter of their offspring receive a copy from both parents. In the past, before any effective medical intervention was possible, most such children died early and unpleasantly from sickle-cell anemia. Today their prognosis is better, but they still suffer a great deal. Their growth is often stunted, and they find physical exertion difficult or impossible, because exercise often precipitates a painful sickle-cell crisis.

Sooner or later in the history of populations, this negative effect counterbalances the benefit of malarial resistance and places a "ceiling" on the frequency of the Hemoglobin-S allele. Exactly what that ceiling is for a particular population depends on the severity of malaria in the region and several other genetic factors. In Senegal today, somewhere between 15 and 20 percent of the population carries the Hemoglobin-S allele. In other parts of Africa where malaria is more prevalent, the frequency of the protective allele can be as high as 40 percent.

This situation demonstrates vividly the strengths, the weaknesses, and the catch-as-catch-can nature of the evolutionary process. The Hemoglobin-S allele is by no means a "perfect" solution to the problem posed by malaria. It was a random mutation in the hemoglobin gene that happened to make life difficult for malaria parasites without destroying hemoglobin's normal function altogether. And while a single copy of the gene protects against malaria, two copies cause sickle-cell anemia, an example of what evolutionary biologists term a compromise and what cynics might liken to a bargain with the devil.

THE MALARIA-HEMOGLOBIN story has many more twists and turns, for Hemoglobin-S isn't our species' only evolutionary response to malaria. Human populations around the world display dozens of genetic malaria deterrents, including twenty different changes to the hemoglobin molecule alone. Several of these conditions can occur simultaneously, and their alleles interact in complex ways, all of which have important repercussions for human health.

The Hemoglobin-S allele, for example, often co-occurs with another mutation that also boosts malaria resistance. This unidentified allele affects the production of still another form of hemoglobin, called Hemoglobin-F because it is found in fetuses and newborns. Normally, Hemoglobin-F production is "switched off" by genetic controls soon after birth, but one or more mutant alleles can somehow delay that off switch. As a result, fetal hemoglobin genes stay turned on, and Hemoglobin-F is made during adult life. People with the resulting mixture of sickle and fetal hemoglobin seem to have the best of both worlds. Malaria parasites don't grow well on either form of the protein, and the mixture prevents the sickling that would occur with Hemoglobin-S alone. Therefore, people who carry two copies of the Hemoglobin-S allele but also produce Hemoglobin-F don't suffer as much from sickle-cell anemia.

Such variations are enormously exciting to clinical researchers like Nagel. "As physicians," he explains, "we are particularly grateful for mutations that give us innate resistance to infection or other diseases. As a matter of fact, our learning about them allows us to treat diseases in individuals who do not have the genetic makeup to defend themselves. So they also give us important tools to deal with disease for everybody."

Nagel is working feverishly to find out how the Hemoglobin-F off switch is delayed. He hopes that his work will ultimately allow doctors to turn on the production of fetal hemoglobin in adults by using the molecular tools you will meet in later chapters. That ability could save the lives of those suffering from sickle-cell anemia, and open new avenues of research aimed at bolstering malaria resistance among people who don't carry the Hemoglobin-S allele.

Again, those of us far from malarial regions might not realize how vital this work is. Mutations involving hemoglobin are powerful, dynamic, and *living* defenses against enemies which are themselves alive; these changes in our genes have fought malaria for millennia, for they enable our species to evolve as parasites evolve. In contrast, insecticides aimed at mosquitoes and drugs to kill malaria parasites are static, inert responses that are rapidly losing their effectiveness. Mosquitoes around the world now carry mutated genes that enable them to tolerate, eliminate, or inactivate our poisons. *Plasmodium* strains in many areas carry other recently altered alleles that enable them to resist not just one or two, but several of our antimalarial drugs.

This evolutionary counterattack has been as successful as it is implacable; during 1992, more than 300 million people were infected with malaria, a rate higher than it had been for decades. We who venture into drug-resistant malarial regions lacking genetic defenses do so increasingly at our own risk. In 1993, a revolutionary malaria vaccine devised by molecular researchers in Colombia was donated to the World Health Organization (WHO). It holds promise to fend off the parasites for a time, although (as you will see in Chapter 5) it will probably need to be updated regularly. In the meantime, tropical populations are likely to need their home-grown malarial deterrents, plus any new ones natural selection happens to concoct, for some time to come.

• • •

THE STORY OF Hemoglobin-S shows how small changes in genes can help species respond to an environmental challenge such as the appearance of a new disease. That story came together over time because all its components—Hemoglobin-S, ordinary hemoglobin, and malaria—have been of sufficient medical importance to attract interest among clinical researchers and funding agencies for many years. Over the past few years, meanwhile, evolutionary biologists have been avidly appropriating molecular tools developed by their more clinically focused colleagues and applying them to questions about the diversity of life that have perplexed evolutionary scholars since Darwin's time.

"The land," noted Darwin in *Voyage of the Beagle,* "is one great wild, untidy, luxuriant hothouse, made by nature for herself."[5] That hothouse is populated by somewhere between 3 million and 40 million species of living things, adapted to environments ranging from tropical rain forests, to arctic mountains, to the depths of the sea. In parts of the tropics, local diversity is overwhelming; Ecuador, a country barely the size of Colorado, hosts more than 700 species of reptiles and amphibians, 1,300 species of birds, and many thousand species of insects.

Yet this living multitude, together with such well known fossils as dinosaurs, saber-toothed tigers, and the like, represents only a fraction of the species that have come and gone since life began. In British Columbia, for example, a single geological formation, the Burgess Shale, preserves a fauna containing more different forms of life than now inhabit the world's oceans. Most of them vanished in the vastness of time, leaving no descendants. Yet this rich and bizarre fauna is but a single snapshot from one page in the family album of life, an album that covers more than 3.5 billion years of earth's history. Conservative estimates suggest that if such an album actually existed, its pages would contain a minimum of 2 *billion* species.

Biologists have struggled to explain this staggering abundance ever since Darwin set sail on the *Beagle.* As recently as the late 1970s many biologists thought they had reached their limits in terms of relating evolutionary change directly to changes in DNA, because precise manipulation of genetic material was either impossible or too complicated, expensive, and time-consuming to permit its use in studies of large populations. Today, however, streamlined and automated molecular tools enable researchers to read DNA's base sequences, study similarities and differences among various alleles, track the movements of genes through populations, and determine the effects of specific alleles on the lives of plants and animals. Few of these efforts have produced the sort of "smoking gun" provided by Hemoglobin-S, but many are making enormous strides in either tracing or explaining evolutionary change.

Several such groups of researchers concentrate their efforts in the Hawaiian archipelago, an ideal location for studying both the fruits and the processes of evolution. The archipelago is young; its newest member, the Big Island of Hawaii, which thrust its head above the waves a mere 400,000 years ago, is still being built by active volcanoes. Although the islands stand in the middle of the Pacific Ocean, far from major land masses, they host hundreds of unusual plant and animal species, many of which are found nowhere else in the world. Where did those organisms come from? How did they evolve?

From a vantage point on a rocky, windswept ridge above a verdant Hawaiian valley, it is easy to understand how Darwin was inspired by his visit to another group of islands, the Galápagos, during the *Beagle*'s voyage. Some years later, he wrote:

> Considering the small size of these islands, we feel the more astonished at the number of their aboriginal beings, and their confined range. Seeing every height crowned with its crater, and the boundaries of most lava-streams still distinct, we are led to believe that within a period geologically recent the unbroken ocean was here spread out. Hence both in space and time, we seem to be brought somewhat near to that great fact—that mystery of mysteries—the first appearance of new beings on this earth.[6]

English naturalist Alfred Wallace, who almost scooped Darwin with independent work on evolutionary theory, was also inspired by island flora and fauna, in his case in the archipelagoes of Indonesia and Malaysia.

Biologists who study Hawaii today enjoy another advantage: the archipelago offers both organisms and researchers an extraordinary diversity of environments concentrated into very small areas. The windward side of the Big Island, for example, is usually swaddled in clouds spawned as moist air climbs the volcanic slopes of Mauna Loa and Mauna Kea. Downwind of the mountains, in contrast, the Kona coast bakes under near-desert conditions. Temperature falls predictably with increasing elevation; the coast is always tropical, yet snow falls on the tallest peaks in winter.

Other environmental conditions vary just as much, and just as sharply. Some lava flows are literally warm to the touch and offer few footholds for plant and animal life. Lava flows in these islands, however, rarely blanket the countryside; instead, they split into glowing tongues that snake toward the sea, sparing areas between them that Hawaiians call *kipukas*. Kipukas, which remain as miniature "islands" of old growth amid newly solidified lava, vary in age from a few years to several centuries and range in size from a few feet to several miles across.

The resulting environmental patchwork impresses even biologists who understand and anticipate it. In less than a mile, hikers can pass from a hot, dry desert into a cool, moist, tree fern forest. Within a few inches, the ground underfoot can change from jagged, shoe-slicing, recently cooled lava to soft, ancient, moss-covered humus. Each section of this environmental quilt is likely to harbor at least some species that are unique, either to the island, or even to a single kipuka. No wonder biologists view Hawaii as one of the world's premier natural laboratories for evolutionary studies.

ONE OF HAWAII'S most spectacular examples of plant evolution is a group of species called the silversword alliance after the charismatic plants whose pointed silver-haired leaves explain their common name. These plants, found only in the Hawaiian archipelago, share enough details of leaf and flower structure to convince botanists that they are closely related. Yet beyond flowers that look like a cross between daisies and sunflowers, the thirty-nine known members of the group are as different as night and day, both in appearance and in the habitats to which they are adapted.

Among the most prominent members of the group is the Maui silversword, whose six-foot spikes of reddish brown flowers tower above a dense cluster of swordlike leaves. That plant inhabits the volcanic crater of Haleakala on Maui, a spot drier than many deserts. Other species, diminutive creeping shrubs with tiny fuzzy leaves, cling to recent lava flows thousands of feet above sea level. Another species, with broad, smooth leaves, grows into tall, branching trees and shrubs that live in or near Kauai's Alakai swamp, arguably the wettest place on earth. Additional members of the group climb like vines.

Where did all the silverswords come from? That's the question a trio of botanists—Gerry Carr of the University of Hawaii, and Rob Robichaux and Bruce Baldwin, both of the University of Arizona—joined forces to study. Originally, they knew, silverswords' ancestors had to come from elsewhere, probably in the form of an errant seed or two. Earlier workers had established that the ancestors of most Hawaiian plants came from Southeast Asia and Australia, borne on favorable winds and currents and "refreshed" along the way by sojourns on islands that punctuate the open sea. But decades ago, naturalist Sherwin Carlquist suggested a radically different origin for the silverswords: a group of North American plants called tarweeds, whose habitat is separated from Hawaii by 2,400 miles of unbroken ocean.

Proof of that relationship was tough to come by. Botanists usually demonstrate close kinship between two plants by crossing them to produce hybrids, because unrelated species rarely cross with one another. But attempts to match members of the silversword alliance with similar-looking California tarweed species proved futile; none of the intuitively obvious crosses took. Researchers quickly realized that a better approach was needed, for crossing each of thirty-nine members of the silversword alliance with each of ninety-nine tarweed species could take a lifetime.

Carr, Robichaux, and Baldwin resolved to bring the tools of molecular biology to bear on this question by analyzing the DNA of silverswords in parallel with studies on their structure and natural history. They and their colleagues compared the DNA of silverswords and tarweeds, looking for enough genetic similarity to indicate close ties between island plants and mainland species.[7] Their work singled out two tarweeds and one silversword as likely close relatives. When the crosses were made they produced healthy hybrid plants. Silverswords *had* come from North American ancestors.

In retrospect, Baldwin notes that the tarweed most closely related to the silverswords has exactly the characteristics one might expect in a plant able to cross the Pacific and colonize volcanic islands. This species was discovered by John Muir in the Sierra Nevada, where it grows only on isolated granite outcrops. Not surprisingly, it is the best of its kin at traveling—or more precisely, at getting itself carried—from place to place. Bristles on its fruits lock tightly into feathers and fur of wide-ranging birds and animals. Undoubtedly one such bird carried a few seeds to the oldest islands in the Hawaiian chain millions of years ago.

After that fateful journey, the plants that sprouted from those seeds would have had a good chance of surviving, because there were few competing plants or plant-eating animals on the infant islands. Sooner or later, one or more of the founding plants produced seeds that carried mutant genes. Many of those mutants would have been lost causes, unable to grow at all. But some could have grown into seedlings that, instead of being suited to growth in dry, rocky environments, could grow in wet spots.

By chance, speculates Baldwin, such a seed might have been carried by wind or animals to a wet place on the island. (Swamps and deserts are quite close to each other in Hawaii.) That new, water-tolerant plant in a wet location, probably almost free of both competition and enemies, would have had a good chance to survive. Generation after generation, the same process of random variation and dispersal to new habitats could have produced numerous separate populations.

Once plants spread to other islands, they would have been isolated from relatives. Little by little, a combination of random genetic changes and natural selection could have caused the genetic texts of isolated populations to diverge. Today, most of these plants are distinct enough in appearance and habitat that botanists recognize them as separate species. Genetically, however, they are nearly as closely related to one another as diverse groups of humans; even the most physically distinct silversword species share DNA similarities of more than 95 percent. Still, many members of the alliance interbreed with one another only rarely, though they may be close neighbors.

Such is the case with two low-growing, silvery-leaved species within the genus *Duboutia,* plants that literally brush leaves with each other as they divide their habitat in the treeless volcanic landscape between Mauna Loa and Mauna Kea. Alternately baked in the sun and shrouded in mist, they grow on a mosaic of lava flows. "The black flow," Rob Robichaux points out, "the one that looks like smooth globs of chocolate pudding, is called Pahoehoe lava and dates from 1935. The red lava, down in all these small kipukas here, is a prehistoric flow that's mostly A'ah lava. It looks more like burnt popcorn."

That intermingling of these flows creates a dividing line between the two *Duboutia* species. "From here," Robichaux announces, waving his hands first toward the sea and then toward the nearby peak, "*Duboutia scabra* goes down into the wettest of the wet forests, and *Duboutia cilliolata* goes way up Mauna Kea into much drier environments. But wherever you get this patchwork of flows, you pick up one or the other in the appropriate place."

The "appropriate place" is an abrupt boundary between new lava and old. The transition zone covers only a fraction of an inch. Yet something about the composition or consistency of the two flows creates a critical, and still enigmatic, ecological boundary between the two species. "*Duboutia scabra,*" Robichaux explains, "is always one of the first colonizers on new flows, a real pioneer. *Duboutia cilliolata* grows only in the older kipukas. We've spent hundreds of hours looking at these plants and we've never found any that switch places. You can find places where the crowns of the two species are touching each other, but their roots are in different flows." What genetically controlled characteristics account for this ecological partitioning? And what combination of genes keeps the plants from interbreeding? Those are just two of the questions that Robichaux and his colleagues hope to answer over time.

Other questions face them everywhere. A few miles away, in a ravine on the eastern slopes of Mauna Kea, a relative of the fuzzy little lava weeds, *Duboutia arborea,* grows to ten feet in height. Scattered up and down that ravine are scores of different growth forms, which appear to represent rare natural hybrids between that species and scruffy *Duboutia cilliolata.* Some of those hybrids are as tall as their treelike parent, while others are scarcely six inches high; some have shiny broad leaves, while others carry leaves that are long, narrow, and fuzzy. Each form seems to manage best in a particular spot: a few thrive on exposed rocks, others deep in the ravine where water is abundant, and still others in the shade of rocky walls. If these hybrids survive long enough, they may become so genetically different from one another that they will no longer interbreed easily. At that point, they will be on the way to forming new species.

Which genes and how many of them account for these dramatic differences among such closely related plants? Are these plants in fact evolving into new species? Baldwin, Carr, and Robichaux, who have spent most of their lives addressing such questions, aren't ready to make predictions. They all agree that this single ravine probably holds enough riddles to occupy them for lifetimes to come.

• • •

ALTHOUGH THE HAWAIIAN researchers' studies examine DNA to trace relationships among species backward in time, and although the Hemoglobin-S story shows how a mutation can enable a population to survive under adversity, neither case can tell us much about a profoundly important evolutionary phenomenon: the reshaping of a species through the evolution of entirely new features that enable it to make its living in radically different ways. One emerging tale of molecular evolution that demonstrates such a change is no further away than the average dinner table or the nearest farm.

The most abundant mammals on earth, and those which humans most depend on for meat and dairy products, are all relatives of cows, sheep, goats, and deer, animals known as ruminants. All ruminants have a unique ability: they can survive on a diet of grass, twigs, and leaves which they digest into useful nutrients. That ability is remarkable because the plant parts ruminants eat can't support most other mammals. Nuts, fruits, vegetables, and berries are packed with energy-rich compounds. But grass, twigs, and leaves are mostly cellulose, which no mammalian stomach enzymes can digest. So how do these animals manage?

When a typical ruminant, such as a cow, feeds, it grinds food into a pulp and swallows it into a stomach with four chambers. In the first chamber, or rumen, pulverized food is mixed with digestive enzymes and a rich culture of bacteria. Those bacteria produce their own cocktail of enzymes, one of which breaks down cellulose. As the bacteria digest the cellulose, they grow and multiply. When the slurry of food and bacteria has been processed sufficiently, it passes into the cow's next stomach chamber. There the cow turns the tables on the bacteria it has nurtured thus far, digesting them and turning them into a supplemental meal. To do so, however, the cow must somehow demolish the tough cell wall that surrounds each bacterium, a feat, like digesting cellulose, which few mammalian enzymes can manage.

The enzyme that ruminants use to do this job is the focus of a molecular puzzle that has occupied David Irwin of the University of Toronto for several years. It turns out that this bacteria-digesting enzyme is nearly, though not completely, identical to *lysozyme,* a common enzyme. Lysozyme is produced by nearly every animal species in tears, saliva, and blood, where its ability to digest bacteria makes it an excellent defense against infection.

Because all animals seem to make lysozyme, Irwin explains, it probably arose very early on in evolution. Given that history and distribution, you might wonder what's so special about finding it in the stomach. But lysozyme in tears, saliva, and blood functions in benign, nearly neutral solutions, while the stomach is full of hydrochloric acid and powerful enzymes. Ordinary lysozyme couldn't possibly function under those conditions for two reasons.

First, normal lysozyme would be digested by stomach enzymes just as most other proteins would be. Second, proteins are sensitive to changes in acidity, as you may have witnessed in a cup of afternoon tea. If you take milk in your tea, you know that the two liquids mix well. But if you add lemon juice to the cup, the milk proteins curdle, changing their structure because of the acidity of lemon juice. The same thing would happen to lysozyme from tears if it were secreted into the stomach.

For lysozyme to work in the stomach, therefore, it had to be changed in ways that protect it while enabling it to work. The process may sound simple, given all the other tasks that evolution seems capable of performing, except for one problem: lysozyme was already tailored to do its vital antibacterial work in the eyes, mouth, and blood. So how could a different version of the gene evolve without jeopardizing its primary function?

What happens, Irwin says, is that every once in a while DNA copying machinery—like a photocopying machine run amok—makes a major slipup and creates extra copies of whole pieces of the genome. When that occurs, the descendants of the organism in which the duplication occurred carry extra copies of the duplicated genes. Sometimes nothing of value happens to those copies. In fact, some of our junk DNA seems to be remnants of old genes that were duplicated and then abandoned. But at other times those copies change and take on radically new functions.

According to Irwin, that's how ruminants' stomach lysozyme evolved. When he examined the lysozyme gene, he discovered that all ruminants carry several copies of it. By comparing the number of copies in living species and working backward through the animals' family trees, he traced the original duplication to somewhere around 55 million years ago, then found a subsequent duplication between 25 and 40 million years ago. After the gene was duplicated, chance and natural selection could "fiddle" with those extra copies with no adverse effects to the animals carrying them. The original copy of the gene could remain unchanged, its product continuing to fight infection in tears and other places.

Ultimately, random changes in base sequences in some of the extra lysozyme gene copies altered the enzyme's chemical composition in a manner that just happened to make it functional in the stomach. In parallel, segments of DNA which turn that modified gene on and off changed in such a way that the altered gene was manufactured by cells lining the stomach rather than in places where normal lysozyme is made.

Clearly, all these cumulative changes didn't happen overnight. "It looks as if the stomach lysozyme initially appeared on the scene 55, 60 million years ago," Irwin suggests. "But it probably wasn't until 25 or 40 million years ago that it actually evolved into an enzyme which was successful in the stomach. So it took 20 to 30 million years of accumulating changes that adapted the lysozyme all the way from a non-stomach enzyme to a very good stomach enzyme."

Along the way, even a partially functional stomach lysozyme would enable animals carrying it to take advantage of a food source that no other large animal could use. Eventually this new lysozyme became so good at doing its job that ruminants could rely exclusively on leaves and grasses. That ability, in turn, had far-reaching evolutionary and ecological consequences. "Ruminants have become the dominant large herbivores on earth," Irwin notes. "They replaced previous herbivores that looked a lot like horses and rhinoceroses, and spread throughout all the continents—Africa, Europe, Asia, the Americas—even before humans came upon the scene."

Ruminants are lucky that their ancestors evolved these new genes, for several other animals that could use them don't seem to have them. This unlikely crew includes the koala, the leaf-eating langur monkey, and one of the only leaf-eating birds in the world, the Amazonian hoatzin. Both hoatzin and langur use lysozyme in their stomachs to digest bacteria, but because they have only one copy of the gene turning out the enzyme, they can't digest bacteria as efficiently as ruminants do. The monkeys, in addition, struck a hard evolutionary bargain for leaf-eating ability; in evolving acid-tolerant lysozyme, they lost the enzyme's standard formula, which means their eyes and mouths might not be so well protected against bacterial infection. Note that none of these animals has been nearly as successful in the global scheme of things as ruminants have been.

The story of the lysozyme gene duplication and evolution isn't an isolated example. Several studies have shown that gene duplication plays an important role in evolutionary adaptation. Certain strains of insecticide-resistant mosquitoes, for example, tolerate particular pesticides because a gene that codes for a vital enzyme has been duplicated several times; the increased production of this enzyme enables the animals to inactivate the pesticide before it does them any harm. According to researchers who have studied this phenomenon, the duplication occurred first in an African mosquito population and spread quickly across the Old World.

Even these cases barely touch on the evolutionary significance of gene duplication. In the past few years, molecular biologists have identified several groups of genes that seem to have evolved from a single ancestral gene which was duplicated and modified repeatedly over millions of years. Such groups are called gene families, and some of them contain as many as 200 genes.

One striking example is the enormous group of genes called the *immunoglobulin gene superfamily*. Although all genes in this family produce proteins with different functions, most of them help cells recognize one another. Some enable communication between cells of the immune system. Others produce proteins that direct growing fibers of the nervous system, helping them determine where they are, where they should be heading, and where they should hook up in the developing brain. Still other related genes turn out proteins that receive and translate chemical messages passing between cells. In short, members of this gene family perform many of the tasks that enable single, independent cells to form multicellular organisms.

Several molecular biologists, working with a wide range of important gene families, have tried to determine when the first main burst of duplication in these genes took place. Their data led Leroy Hood of the University of Washington and several colleagues to speculate that the original duplication occurred somewhere between 600 and 700 million years ago, which seems to have been a critical point in evolutionary time. Previously, nearly 3 billion years had produced only single-celled forms of life. We say "only" because those organisms did evolve the complex molecular machinery, including the original DNA-based genome that makes life possible.

According to one hypothesis, that original genome, which had been tuned, refined, and elaborated upon for nearly 3 billion years, was suddenly duplicated, providing "extra" copies of essential genes on which mutation, chance, and natural selection could operate with far fewer constraints than previously. That primordial duplication might have "bootstrapped" our single-celled ancestors up to a level of genetic complexity that enabled all of multicellular life to evolve. In no time, a myriad of different life forms appeared and scurried off along different paths, "trying out" different ways of being multicellular and producing the Burgess Shale and other peculiar early faunas. A little while later, geologically speaking of course, a small band of primates wandered out of Africa. The rest, as they say, is history.

• • •

YOU'VE SEEN so far that evolution means "change," at the level of both molecules and organisms. The persistence of life through the eons often, though not always, requires change, although that change need not be constant and can be neither directed nor purposeful. Change is often vital because environmental conditions with which organisms must cope are in constant flux. When a new form of malaria acquired the ability to attack *Homo sapiens,* for example, that change in the order of life spelled opportunity for the parasite and a deadly challenge for humans. If such challenges are severe enough, the option is simple: change or die—or, in biological terms, evolve or face extinction.

That's why many biologists feel that a population with substantial variation in its DNA has the best chance of surviving over time. Why? Because the more genetic variation exists in a population, the greater the chance that at least some individuals will carry alleles that enable them to survive a new challenge. This reasoning intuitively seems valid; genetic diversity seems to be a valuable asset. But are there any solid data indicating that genetic diversity is truly vital to species survival? If so, what happens to a species that lacks genetic diversity?

For answers, we return to the island of Hawaii, where one private residence sports an artificial lake surrounded by a minefield of traps to snare mongooses and rats, enemies of ground-breeding birds. Strutting imperiously around are the reasons for both the lake and the traps: the handsome geese called the nene (pronounced nay-nay). This endangered species, the state bird of Hawaii, is the subject of studies on genetic diversity that may serve as a model for endangered species of all types around the world.

Fossil evidence tells us that nenes were once abundant on all the main Hawaiian islands. Polynesians, however, included the birds in their cuisine, and by the time Captain James Cook arrived, only 50,000 to 100,000 remained, and they were found only on Hawaii and Maui. Soon whalers plying the Pacific began killing nenes, too, salting them and taking them out to sea in huge numbers. Farming spread, destroying the birds' habitat, and the introduced mongooses and rats devoured eggs and young. Nene populations plummeted.

Roy Blackshear, nephew of Hawaiian conservationist Herbert Shipman, explains why nenes almost went extinct, but didn't. "Uncle Herbert was a real conservationist," Blackshear recalls. "He was given the first pair about 1918. Even at that time he knew that they were rare birds. He had this pair that was given to him by a Mrs. Hind over in Kona, and he started raising them. When I was a kid, I remember we had about thirty-eight birds over here in 1938. Then the 1946 tidal wave came through and took out about half the population. After the tidal wave, he had something like twenty, twenty-one birds. But he told me, 'We have to do something with these birds. Because as far as I know, these are the last nenes in the world.' "

Uncle Herbert wasn't quite right, but he was close to the truth—and the truth soon got closer to his statement. In 1936, when there were probably only several hundred nenes left in Hawaii, Britain's earl of Darby took a few males and females to his home in England, bred them, and distributed the offspring to zoos throughout Europe. But then came the Second World War. Those birds which didn't succumb to bullets and bombs are rumored to have been eaten by the Nazis. In 1950, just seventeen nenes were counted on all the islands, and a total of no more than thirty or so probably existed.

The birds were experiencing what biologists call a population bottleneck, a reduction of numbers almost to the point of extinction. Shipman's efforts during those years undoubtedly saved the nene from extinction. In addition to maintaining his own flock, he contributed three birds to the state captive-breeding program on the saddle road between Mauna Loa and Mauna Kea, a spot selected because it was far away from most mongooses and rats. The six birds that began that program—three of Uncle Herbert's, two taken from the wild, and one from the Honolulu Zoo—are the parents of most, if not all, living nenes.

Today there are several hundred nenes, about half in captive-breeding programs and half in the wild. That number seems large, but efforts to reestablish birds in the wild have met with little success. "Nenes in most locations have not been sustaining their populations," admits Rob Fleischer of the National Zoo in Washington. "They haven't been reproducing, and survivorship of reintroduced birds has been fairly low. This is very unfortunate. They're doing fine in captivity, where we're able to produce a lot. But in the wild there are still problems that need to be fixed."

The birds' failure to thrive continues despite successful efforts to protect eggs from mongoose and rat predation. Eggs don't seem to be hatching, and goslings don't seem to be surviving, an indication that something may be wrong with the geese themselves. What could be amiss? Several biologists fear that the population bottleneck during the 1930s and 1940s may have eliminated much of the population's naturally occurring genetic diversity. Because most birds alive today are related to the few that Shipman saved, they are all close relatives. And that could be causing difficulties.

You've probably heard of highly inbred dog breeds known for problems of one sort or another. Some, like golden retrievers, are prone to hip dysplasia and other debilitating ailments. There is also what animal breeders call inbreeding depression, a syndrome that includes low fertility, reproductive failure, and general failure to thrive.

In the past, conservation biologists could only guess what might be happening to these birds. But molecular techniques now enable them to study endangered animals' genetic makeup in ways that may pinpoint problems and help with solutions. Fleischer and his colleagues are taking blood samples from wild and captive nenes and producing a type of genetic fingerprint for each individual. They then use these data both to evaluate general genetic diversity in the living population and to investigate the blood lines of as many living birds as possible.

Results to date have not been encouraging. "So far," says Fleischer, shrugging, "our analyses haven't identified any variation in wild flocks that isn't present in captive flocks. If we look at DNA fingerprints of captive and wild birds, we see that most of them have a lot of genes in common. This is more or less expected, because we know that most birds in the wild today were probably derived from captive birds or were captive-release birds themselves. But their fingerprints show much less variation than the DNA fingerprints from populations of normal, wild, nonendangered species. This indicates that the nene as a whole is lacking variation compared with other species."

Trying to determine just how serious the situation is, biologist Liz Rave is working with Fleischer, taking DNA samples from banded, free-ranging nenes. "What we're hoping to find," Rave explains, "is that some of the original wild birds have left descendants throughout the years. We want to find out if those descendants have any unique genes that aren't represented in the captive colonies. If that's true, we want to get some of those birds or the offspring of those birds and bring them back into the captive group to increase the genetic variation within the colony."

As they gather these data, Fleischer and Rave also seek answers to a larger riddle. They suspect that nenes lost most of their genetic variation during the population bottleneck of the 1930s. Is that really the case? In the past, there would have been no way to answer that question. But today the combination of DNA fingerprinting and polymerase chain reaction (PCR) can do the trick. While Rave is out catching live birds, Fleischer is taking DNA samples from dead ones. Using the magic of PCR, he is amplifying tiny, degraded pieces of DNA extracted from old museum specimens, some of which were collected almost a century ago. He then produces fingerprints from those samples and compares them with those taken from the DNA of living birds.

The results to date, while scientifically gratifying, don't augur well for nenes. "We have so far found no variation among current nene populations in the sequence of a particular gene," Fleischer reports. "But in a few of the museum samples we have found some variation. What this means is that the bottleneck *did* reduce genetic variation. We don't have a large enough sample yet to know exactly the extent of it, but we soon will. The important thing we have found is that the variation that was present in the past is not there now."

Why is that finding so troubling? There are no data linking lack of diversity in modern nene populations with birds' failure to thrive in the wild. But there are such data for another species, the endangered cheetah, which show that loss of diversity can have serious consequences. Cheetahs, which once ranged over much of the world, are now restricted to a few isolated populations in Africa. The species as a whole seems to have experienced particularly severe population bottlenecks during the nineteenth century.

A study at the National Cancer Institute that examined 200 DNA sequences from fifty-five cheetahs found variation in fewer than one percent of the genetic locations examined. By way of comparison, equivalent human studies typically show variation in roughly 32 percent of similar sites, while other wild animal species typically show variation as high as 36 percent. What has this lack of diversity meant for cheetahs? Most cheetahs in captivity are as closely related to one another as an inbred strain of mice would be after ten to twenty generations of matings between brothers and sisters. As a result, these cats usually accept skin grafts from one another, because their tissues are so alike that individuals' immune systems don't recognize skin from other cheetahs as foreign. Despite intensive efforts to breed these animals in captivity, successful births are rare. Sperm counts in cheetahs run roughly one-tenth those of domestic cats, and more than twice as many cheetah sperm show such abnormalities as bent tails and malformed heads.

The predicament of wild cheetahs is particularly distressing, because the introduction of domestic cats to Maasai villages has exposed wild cats to new feline diseases. Because the genes of wild cheetahs are so similar to one another, a disease that can evade one animal's immune system is likely to wreak havoc on the whole population. An epidemic of feline infectious peritonitis in the United States provides an ominous precedent. While that disease is fatal to only about one percent of domestic cats, it killed 60 percent of the cheetahs in a previously successful breeding program at Oregon's Wildlife Safari Park. The park's lions, which carried more normal levels of genetic variation, were also exposed, but they scarcely showed symptoms.

The nene and cheetah studies have major repercussions for the management of threatened and endangered species everywhere. A host of questions, similar to those Fleischer and Rave are asking about the nene, apply to nearly every plant and animal species threatened by human activity: How likely is it that a population bottleneck will restrict genetic diversity? Is there any measure of genetic diversity that can serve as a predictor of species viability? Just how small does a population have to get before its genetic constitution is permanently impaired?

Fleischer points out that a lack of genetic diversity may underlie the extinction of more species than we suspect. The important detail to remember, he notes, is that genetic diversity can be lost very rapidly in small populations. Once the normal complement of alternative alleles is lost, it is most difficult to recover. How long does such a recovery take? "It depends on mutation rates," he answers with a sigh, "which are notoriously slow in birds. So it may be thousands of generations before the nenes recover the genetic variation they lost in a few decades."

Hovering over populations with low genetic diversity, whatever their size, is the threat of what Fleischer calls a *genetic vortex*, a powerful whirlpool that can drag a species down to extinction suddenly and unexpectedly. "Once the species or the population no longer has enough variation for adaptation, individuals tend to be closely related to one another. There are high levels of inbreeding and inbreeding depression, and along with random factors, this can cause species to drop down to a few individuals and then to none."

Understanding, rather than just observing, population bottlenecks and genetic diversity is important, for the story is not always the same. "If you look at all the species that have been reintroduced to the wild," Fleischer points out, "you'll find that in some cases where only a few individuals were introduced, the population did not go extinct, but instead actually grew and flourished. Those success stories give us hope that the nenes will recover."

IN CLOSING, it's worth noting that *Homo sapiens*, like most other animal species, has substantial genetic diversity. Until the advent of modern transportation, human populations scattered around the globe had little contact with one another, so they rarely exchanged genes. Over the course of our species' evolutionary history, random changes in the genome and natural selection have produced a host of different traits, including visible physical features, such as skin color and eye shape, and invisible physiological traits, such as Hemoglobin-S and other genetic malaria deterrents. Luckily for all of us, there is a great deal more to that diversity, which we are just beginning to fathom.

One study in San Francisco, for example, has turned up preliminary evidence that some individuals may carry genes that either protect them from the *human immunodeficiency virus* (HIV) or at least hinder the progress of AIDS after infection occurs. If these studies are confirmed, and if those genes can be identified, they may open new doors, not only in the treatment of AIDS, but in our understanding of how the body fights other viral diseases as well.

Unfortunately, this vital biological feature of our species, which all of us should celebrate and be thankful for, has been misunderstood and misused in ways that have caused (and continue to cause) much pain and suffering. Recent years have seen revivals of terms such as "racial purity" and "ethnic cleansing." However, molecular studies of human genetics show that these terms are as scientifically meaningless as they are morally repugnant.

While genetic diversity distinguishes each individual human from every other person in the world, our shared genetic heritage unites us as a species. Most important, although certain alleles are more prominent than others in particular ethnic groups, most genes are *not* segregated among the traditional racial classifications of Caucasoid, Mongoloid, and Negroid. In other words, the concept of "race," often invested with disproportionate social significance, is meaningless on a deep biological level.

That is true in part because explorers and merchants have spread their genes around the world over the last several centuries. Analyses of DNA sequences adjacent to the hemoglobin gene, for example, can distinguish between Hemoglobin-S alleles that arose in Africa and those which arose in other populations. These studies reveal that the mutations which produced modern Hemoglobin-S alleles arose a minimum of three times in different populations in Africa and at least once in Asia. By identifying specific alleles, researchers can also determine from which founding population a given person's Hemoglobin-S allele originated. Intriguingly, these analyses reveal that alleles which clearly appeared in black African populations crop up in light-skinned people from Italy, Greece, Spain, and Portugal. Conversely, the few alleles that affect physical features typically associated with an individual's perceived race—features such as skin color or eye shape—sometimes do and sometimes do not correlate with other genetic variations.

Population genetics makes another point clear as well; more than 85 percent of human diversity appears between individuals *within* any single population. The sorts of variations that are often found among ethnic groups, like differences in genetically influenced responses to drugs and in various types of inherited diseases, account for only about 8 percent of total human genetic variation. That leaves little more than 6 percent of existing variation, which happens to include alleles for visible characteristics such as skin color, between so-called racial groups. "There could possibly be more genetic differences between myself and my wife," asserts one geneticist, referring to a spouse of his own ethnic group, "than there are between me and a Kalahari bushman."

According to geneticist J. Steven Jones of University College, London, the genetic differences between Africans and Europeans are no greater than those between people from different countries within Europe or within Africa. "Individuals," Jones emphasizes, "not nations, and not races, are the main repository of human variation." Mary Claire King, Berkeley geneticist and passionate advocate for the socially responsible use of genetics, concurs. "The richness of human diversity," she notes, "has the splendid social effect of completely debunking racial mythology. One learns that race is a social construct, ultimately, and not a genetic one."

To those of us who agree, genetic diversity offers a rare example of scientific data that address questions as broad as the history of life and as focused as human brotherhood. Genetic variation is both a fact of life and an essential part of survival, for humans as for other species. It represents an invaluable and irreplaceable heritage that we can be proud to pass on to our descendants. Vive la différence!

3

Smart Genes

Ye have made your way from worm to man,
and much within you is still worm.

Friedrich Nietzsche

In an ornate church sandwiched between a convenience store and a shoe repair shop in Manhattan, an infant girl is anointed with oil and sprinkled with water from a ceremonial basin. The white-robed priest blesses her by name: Maria. Not far away, in a living room on the Upper West Side, an eight-month-old boy wails as a ritual knife deftly removes the tip of his foreskin. The rabbi in attendance welcomes Daniel into the company of his fellows.

Baptism and circumcision, rituals with counterparts in many cultures, consecrate the miracle of life and honor tradition. While initiating infants into society, the ceremonies offer parents and grandparents the chance to laugh and tease one another playfully as they examine babies' faces for physical signs of kinship. "Those are Theresa's eyes, I tell you!" insists Maria's maternal grandfather, Tony. "That's his father's nose, poor guy!" quips Daniel's paternal grandmother, Sarah.

Many traits they notice are controlled by simple genes whose workings seem easy to grasp; one length of DNA directs the assembly of a dark compound, and a baby's irises turn out brown. Another gene operates elsewhere in a child's body, and his hair is tinted blond.

But even those "simple" traits, under close observation, raise one of the central mysteries in molecular biology: How do genes "know" when and where to generate their products? The act of love that produced Maria endowed a single fertilized egg cell with a unique complement of genes. Each time that cell divided to produce more cells, it copied its chromosomes faithfully. As a result, all Maria's trillions of cells carry the same DNA, including the genes that give her dark brown eyes. But if every cell carries the genetic code for making brown pigment, why is it made only in parts of her eyes? Why aren't her lips brown as well? Or her ears?

Those same simple questions apply to each of the 100,000 genes in every one of our more than 250 specialized types of body tissue. Genes that work overtime to fill Daniel's red blood cells with hemoglobin don't "turn on" in the neurons of his brain. The lengths of DNA that make digestive enzymes turn out their products in the lining of Maria's stomach, but don't operate in the muscles of her big toe.

Determining *where* genes are activated is only half the mystery; each gene must be turned on not only in the right place, but at the right stage of life and at the right "volume." Some basic "housekeeping" genes operate in nearly all cells almost all the time. Other genes are turned on in only a few cells, for no more than a few hours, at one critical period in an individual's life, and are then turned off permanently. Some genes create vanishingly small amounts of their products, while others, such as those which produce milk proteins during lactation, churn out ounces. The result is a symphony of molecular activity which begins at the moment of conception and continues until death. That genetic composition, more complex than anything composed by Beethoven or Wagner, has dramatically varied movements that correspond to life's sequential stages.

While Maria was in the womb, the tempo was allegro vivace; the theme was growth and differentiation. The original fertilized egg divided to form two, then four, then eight, identical cells. Each of those eight cells had at its disposal nearly all the genetic information it carried; because each cell could access virtually any gene in its nucleus, it had the ability to produce heart muscle, skin, blood cells, or any of the body's other specialized tissues. In fact, any one of the eight cells, if separated from its neighbors carefully and coddled appropriately, could have produced an entire new embryo that would have been Maria's twin sister.

Instead, as those identical cells divided again and again, different genes were turned on and off in each of them. Some on/off switches were temporary, but many important ones were permanent; the progenitors of brain cells lost the ability to make hemoglobin, while cell lines that would later produce red blood cells lost their ability to carry nervous impulses. Each developing cell still carried exactly the same set of molecular instructions—the same library of genetic information—as every other cell in Maria's body. But as the embryo grew, certain specific volumes of genetic text were opened in each type of growing cell, while other volumes—whole shelves and stacks of volumes, in fact—were closed for life.

The farther along diverging paths the various types of cells traveled, the more constrained their options became. Day after day, those cells divided, migrated, transformed into radically different shapes, and began to perform wildly different tasks; nerve cells carried impulses that would become thoughts, liver cells disarmed poisons, and fat cells stored food for lean periods. By the time Maria was born, most of her cells had been permanently programmed for life; liver cells and fat cells would never think, while brain cells would never detoxify alcohol, as liver cells do.

The molecular tempo of life following birth remains allegro. While Maria cries at the touch of cold water, her eyes, ears, and the cognitive centers in her brain forge connections with one another as nerve cells link up into networks communicating through chemical and electrical signals. Neurons that connect properly will thrive—and probably live for nearly as long as Maria does. Many other brain cells will simply die, joining a host of their fellows that perish as a normal part of development. As Daniel wails, cells throughout his body grow and divide rapidly, extending his arms and legs, expanding his heart and liver, and producing larger lungs. Both children's bodies will continue to grow until they reach adult size in just under two decades. And although the fates of their cells are determined by then, they retain enough flexibility to heal the body if injury occurs; broken bones knit almost seamlessly if properly set, while most cuts and scratches heal without a trace.

Later on, during adult life, the tempo of cellular activity slows to a more graceful adagio, and patterns of gene activity change in ways determined by each individual's age, sex, and personal genetic makeup. Daniel's parents, for example, are no longer outgrowing their clothes every six months; their developmental theme is no longer growth, but maintenance. Some genes in each type of cell slow down; others are turned off. Cells in most of their organs divide just often enough to replace parts that are worn or damaged; others no longer divide at all. Their ability to repair bodily damage has dropped somewhat; broken bones are more serious, and skin injuries more often leave scars.

But certain tissues, such as those which make up the skin and the lining of the intestine, respond to wear and tear by replenishing themselves throughout life. They can do so because they contain a remarkable class of cells, called *stem cells,* which retain some of the genetic flexibility other cells lose as they mature. Throughout life, for example, stem cells in bone marrow keep dividing and differentiating, and they can give rise to any type of blood cell in the body, depending on which hormones and growth factors they are exposed to as they mature.

Other genes in mature cells turn on and off in response to a wide range of specific cues. Following the birth of Maria, many of her mother's cells perform tasks they have never done before. While the cells of her uterus recover from pregnancy, the cells of her milk glands produce and secrete prodigious amounts of proteins, sugars, and other compounds that nourish Maria's rapid growth. All that milk-producing activity is controlled by genes that have been present since conception and will endure until death, but are turned on only when primed by pregnancy and childbirth and encouraged by suckling.

The proud grandparents in both families, meanwhile, age gracefully as their genetic orchestration shifts into a quieter, more stately mode. Some genes that have worked hard for decades are slowing down or changing their activity; over the past twenty years Sarah's hair has turned a silvery gray. Other genes have accumulated dangerous genetic errors over time; Tony has had a prostate cancer removed. Still other genes have simply turned off; since Sarah passed menopause, her reproductive hormones no longer cycle as her daughter's do—and as Maria's will do in a little more than a decade.

BIOLOGISTS HAVE LONG wondered how this genetic orchestration controls growth, differentiation, and maturation. Around the turn of the century, German embryologist Hans Spemann showed that the nucleus of an egg was necessary for it to develop properly. Spemann wrapped a thread around the middle of a frog's egg and tightened the loop, squeezed the nucleus into one half of the cell, but stopped short of cutting the egg in two. Only the half of the egg containing the nucleus divided normally, indicating that something inside that structure was somehow responsible for directing its growth and development. During the 1950s, more elaborate experiments showed that a nucleus transplanted from a partially differentiated embryonic cell could successfully replace an egg cell nucleus that had been surgically removed. Converging insights from several fields identified the critical "something" within the nucleus as DNA.

Then during the 1960s and 1970s, still more sophisticated experiments showed that even the nucleus of a cell removed from the lining of a tadpole's intestine could direct a frog egg to produce an entire tadpole.

Together with the findings of several other research teams, this work suggested that the nuclei of adult cells contained all the genes needed to produce an entire organism. It seemed that although many genes were turned off during development, the information they contained (with a few exceptions) remained intact.

Still, the nature of genetic interactions during development remained a mystery, a "black box" whose inner workings were hidden from view. By the late 1970s, adaptable molecular tools such as restriction enzymes and PCR had enabled researchers to isolate genes from any organism and to read the sequence of bases carried by those genes. But it soon became clear that simply identifying and sequencing genes would not explain how genetic instructions were turned on and off. Instead, researchers needed to perform more sophisticated versions of nuclear transplantation experiments; they needed to pick up, manipulate, and transplant individual genes, to "watch" those genes as they produced their messenger RNA and ultimately their protein products, and to connect gene activity with changes they could identify in cells and tissues. Only in that way could they ultimately peer into the black box of genetic interactions and investigate how genes work together.

This undertaking is a challenge equaled in importance and difficulty only by the effort to understand the brain itself. The potential rewards are staggering: treatment or prevention of illnesses ranging from some birth defects to cancer, and the ability to extend the human life span by years or even decades. Sufficient skill in manipulating blood stem cells, for example, could lead to cures for leukemia, several types of anemia, and certain autoimmune diseases.

The human genome, however, is enormous and difficult to probe; the oft-quoted estimates of 100,000 genes are relatively arbitrary, and may well understate our genetic complexity. Our bodies also contain so many cells that we really can't count them; biologists toss around numbers like 60 trillion or 100 million million without really knowing how many there are. Many researchers have therefore turned to worms, fruit flies, fishes, and mice, simpler animals that enable whole classes of experiments that can't be performed on humans or dogs. The nematode worm *Caenorhabditis elegans* is a case in point. Its body contains barely 1,000 cells, and its genome is slightly less than one-thirtieth the size of our own. Partly because of the worm genome's small size, and partly because geneticists around the world have been working with it for years, it is the only animal genome for which a complete physical map has been completed. The next stage of the worm genome project—sequencing, or reading its genetic code base by base—is well under way.

Nematodes have other advantages, too, according to Robert Horvitz, a developmental biologist at MIT. "These animals are transparent, so you can see through them," Horvitz explains. "In fact, you can see every cell in their bodies. And it turns out that every animal of a given genetic makeup has the same number of cells in almost all the same places." That simplicity and predictability has enabled Horvitz and his colleagues to accomplish a feat which is thus far inconceivable in larger organisms: they have been able to trace the developmental path taken by every cell in the animal, from egg to mature adult. "This worm," Horvitz asserts with obvious pride, "is the only animal for which every cell in the body is identified, and the developmental origin of every one of those cells has also been identified. So we know where every cell comes from and how each is related to the others."

Armed with that detailed developmental map, researchers can search for genes that control the various pathways along which cells grow and change. Certainly the end results of gene activity are dramatically different between human and worm. But thanks to evolutionary ties, many molecular mechanisms of development—the strategies and techniques of turning genes on and off—are common, not only to *C. elegans,* fruit flies, mice, and humans, but to remotely distant bacteria as well.

One unifying theme in development, and one of the most central discoveries in all of molecular biology, is the fact that certain genes are more influential than others. That's because genes are organized in a complicated molecular pecking order. At the bottom of the genetic hierarchy are genes consigned to specific, often repetitive tasks: making hemoglobin, growing hair, or producing digestive enzymes. Above these molecular workhorses are several ranks of "regulator" genes that turn them on and off, switching off the fetal hemoglobin gene during infancy, for example, and turning on the allele that functions during adult life. Above both the workhorses and the genetic equivalent of middle managers are a series of master control genes whose decisions affect dozens, or even hundreds, of their subordinates. Some of these master controls are so vital that mutations which inactivate them are lethal early in embryonic life. But others, though important, can malfunction without being fatal. The consequences of mutations in such master control genes, or others farther down in their chain of command, can be astonishing, as a few exceptional people can attest.

• • •

You could easily walk past Marilyn Saville on the street and not notice anything unusual about this neatly dressed middle-aged woman with short graying hair. You could have dinner with her without suspecting that anything exceptional had gone awry in her life. In conversation you might discover that she has a passion for old cars. You might pick up hints that she passed, years ago, through a dark and unhappy period with which she has since come to terms. But you might not ever learn more—unless, that is, you happened to bring up the topic of swimming.

Marilyn Saville was a good swimmer in high school. An excellent swimmer, in fact. And it was an incident in a swimming pool that first attracted attention to her unusual condition. "I was a good candidate for the Olympics," she remembers. "I had been told that by the spring I would be informed for sure whether I was on the Olympic team. Then I was in a swimming meet and had a small accident."

That minor accident shattered Saville's world, although she would not understand why until years later. Temporarily incapacitated by pain in her abdomen, she was rescued from the bottom of the pool and taken to a hospital. There her mother mentioned to the doctor that although Marilyn was seventeen, she had not yet begun to menstruate. For that reason, while the surgeon repaired what turned out to be a hernia, he also performed some exploratory surgery. Thinking little of this minor mishap, Marilyn looked forward to resuming her training.

About three months later, her father sternly called her upstairs for a talk. Between his embarrassment and her confusion, the ensuing conversation conveyed little information, but Saville remembers it clearly. "He said, 'Marilyn, I have something to tell you. The doctor has just told us that you were born missing some parts. You will probably never be able to have any children because you're missing that . . . you know . . . that . . . uterus. Now don't tell anybody about this. You shouldn't, really. We'll keep this to ourselves.'"

That, according to Saville, was her first indication that she was somehow different from other women. But other indications weren't long in coming, in the form of subtly different treatment by her coaches and fellow team members. "When I got back into swimming practice," she remembers sadly, "they treated me very distantly. From that time on it was 'Get lots of rest, and don't push. It's okay, Marilyn. We're really interested in your well-being.' Gradually they distanced me, until for some reason, which no one would ever speak of, I was out. I never understood."

Her situation went from bad to worse. Just before Saville was married, her father had a private "man-to-man" talk with her fiancé. At first the fellow denied the reality of something he didn't quite understand, but later the secret ate away at him like a cancer. "He treated me openly as a freak," Saville reports. "That's how he described me, that I was just a freak. That I was very lucky to have him because no one else would ever, ever have me. If they knew, they would be repulsed. So I should be grateful and thankful that he would tolerate having me around."

Saville remained married to that man for almost thirty years, even though he became violent and beat her so severely that she was hospitalized several times. In addition to the physical trauma, she was terribly lonely. "I was very afraid," she recalls. "Not being able to talk to people. Not being able to say, 'What's the matter? I'm the same person I used to be.' They didn't understand, so they just withdrew and left me."

What was the mystery that caused Saville so much pain and suffering? What was so terrible about her that no one ever fully explained it until she was twenty-eight years old? Her story, as it came to light, is as biologically interesting as it was socially tragic. For Saville carries a defect in a single gene, a particularly powerful regulator gene with a major role in the chain of events controlling sexual development.

Most of the time, an embryo that receives an X chromosome from both parents is born female. An embryo that receives an X chromosome from its mother and a Y chromosome from its father is born male. But though Saville's cells carry the "male" or Y chromosome, she is psychologically, socially, and to all outward appearances a normal, healthy female. Because of a single genetic defect, her outwardly female body not only lacks female reproductive organs but carries in their place a pair of testes hidden inside her abdomen, near where her ovaries might have been.

As unusual as Saville's situation may sound, it is not especially uncommon; an estimated one in 500 people are affected by similar forms of genetically based sex confusion. Human sexuality in its broadest sense encompasses a host of biological, cultural, and psychological factors, yet biological sex in the narrowest sense is powerfully influenced by surprisingly few genes. If any of those genes malfunction during critical periods in life, they produce individuals that blur the distinction between male and female. To understand Saville's condition, and to appreciate what genes do—and do not—have to do with her plight, we must look closely at how the sex of an embryo is determined.

Our word *sex* derives, interestingly enough, from the Latin *secare*, which means to cut or divide an object that was once whole. Its usage probably dates back to Greek myths in which Zeus cut in half a race of primordial beings that originally had attributes of both male and female. That myth has a surprisingly close correlate in embryonic development, where, for the first six weeks after fertilization, normal male (XY) and female (XX) embryos are structurally and biochemically identical. In what will eventually become the lower abdomen lie identical clumps of cells that will soon form either testes or ovaries. The fascinating point is that these clumps of cells are capable of making either male or female gonads. As a result, the embryo can become either a man or a woman, or, in fact, a hermaphrodite, depending on the instructions those cells receive.

What happens next depends on the genetic instructions carried by the fetus. In a normal XY embryo, one set of genes in those clumps of cells turns on and testes form. The infant testes secrete the hormone testosterone, a chemical messenger that travels through the body and turns on certain genes in several different tissues. The result is the growth of the small but fully formed sexual organs of a newborn male. Years later, during puberty, a great deal more testosterone is made. At that time, many genes in several tissues are affected, producing secondary sexual characteristics including beards, body hair, a deepening of the voice, and a tendency to grow somewhat taller and accumulate more muscle mass.[1]

In the XX embryo, on the other hand, a different set of genes turns on in those clumps of cells. The clumps themselves turn into ovaries, and the external female genitals are formed by exactly the same tissues that form the male external sex organs in the XY fetus. When such an individual reaches puberty, increased output of female sex hormones turns on sets of genes that cause breasts to enlarge and produce the other body features characteristic of a mature woman.

The critical role of the testis in this process was uncovered by Alfred Jost, a geneticist in Nazi-occupied France who worked under extraordinarily adverse conditions. Jost experimented with rabbits, and at times, the only way he could keep his animals alive was to take them home with him at night and sleep with them to keep them warm. His efforts were not in vain. Jost found that if he removed the testis of an XY rabbit embryo early in its development, that embryo would develop into a female. This led him to conclude that the formation of the testes is a critical decision point in determining sex in mammals: an embryo will become a female, regardless of its genetic sex, unless testes are present to produce hormones that push development into the male mode.

But what determined whether or not those early clumps of cells turned into a testis? The first clues came from studies of individuals with unusual combinations of sex chromosomes. One group of unusually short women, for example, have what is called Turner's syndrome; they carry only a single sex chromosome—a lone X. Thus, a single X chromosome seemed sufficient to produce a female. In contrast, another group of chromosomally unusual individuals, apparently males, carry three sex chromosomes—XXY. In these cases, a single Y chromosome was enough to produce a male, even in the presence of two X's. It therefore seemed to researchers that the presence of a Y chromosome, rather than the number of X chromosomes, somehow caused testes to form, creating a male. Scientists began talking about a hypothetical piece of DNA on the Y chromosome that they called the testis-determining factor.

That hypothesis seemed reasonable until the late 1970s, when one research team turned up several individuals who, although they had two X chromosomes, were male. How could that be? All current hypotheses assumed that an individual had to carry a Y chromosome in order to be male. The apparent contradiction was resolved in the early 1980s, when researchers applied their ability to sequence and identify stretches of DNA to investigations of the X chromosomes carried by those unusual males. What they found was both gratifying and unexpected: one of the X chromosomes found in XX males carried recognizable chunks of DNA that are normally found only on Y chromosomes. Somehow, through a most unusual episode of chromosomal rearrangement, a small fragment of a Y chromosome had been inserted into an X.

Geneticists interested in sex determination were naturally curious about this discovery. It seemed that only one tiny piece of the Y chromosome, perhaps only a single gene, was enough to make the fetus male. The rest of the Y chromosome was unimportant, from the standpoint of sex, at any rate. According to this hypothesis, XY women would be missing the crucial gene, while XX men would carry it. If researchers could identify the specific piece of Y chromosome that is present in XX males and missing in XY females, they could find "the gene that makes a man." The hunt for the testis-determining factor was on.

The only stumbling block was the need to identify and contact enough XX men and XY women to permit the necessary studies. Because these individuals are physically normal and healthy men and women, however, neither they nor their physicians would necessarily be aware of their unusual genetic situation. Then David Page of MIT came up with the idea of checking the DNA of people who sought help at fertility clinics. He reasoned that at least some of these otherwise normal XX males and XY females would be unable to produce egg or sperm and therefore be incapable of having children.

That hunch proved correct, and soon several laboratories were racing to find not simply the sizable chunk of DNA that contained the master control for sex, but the actual gene itself. Page's group incorrectly thought they had found it in 1988. As it turned out, they were just a little off target; the piece of DNA they identified was close to what they were looking for, but it wasn't the gene itself. Another likely candidate gene was found in 1991 by a team headed by Peter Goodfellow of the Imperial Cancer Research Fund in London.

To determine whether or not Goodfellow's gene was *the* testis-determining factor, researchers gathered two types of evidence. First, they used a series of technical tricks and an understanding of the genetic code to determine when and where in embryonic life this gene turned on. Once they isolated the gene and made copies of it, they attached a radioactive "tag" to it, so that they could pinpoint its location easily. Then, separating the strands of the gene's double helix, they placed the resulting single strands on top of mouse embryos at various stages of development. After incubating several embryos so treated for different lengths of time, they placed a piece of photographic film on top of each one. That film, they knew, would be exposed by the radioactively labeled DNA wherever it had accumulated within the embryo.

What were they after? The researchers knew that the base sequence of this tagged DNA would carry a sequence of bases complementary to the sequence of messenger RNA produced by the copies of the gene operating within the embryo. Because of that complementarity, the labeled DNA would stick to any strands of that particular messenger RNA with which it came in contact. But messenger RNA produced by the suspect gene wouldn't be present in all cells all the time; it would be made only in the locations and at the moment in time when the gene was activated. Thus, if the radioactively labeled DNA attached to a part of an embryo of a certain age, it would pinpoint the time and place the gene was turned on.

Remarkably, they proved that their candidate gene actively produces its protein only in the clumps of embryonic tissue that produce gonads, and only for about two days—just before the testis is formed. "It's remarkable," Goodfellow admits, "that we're talking about a few thousand cells for perhaps forty-eight hours during development. That's all it takes to convert the female body form into the male body form."

The second, and most striking, proof of the gene's identity came when the researchers successfully inserted it into female (XX) mouse embryos—and transformed them into male mice! There could then be little doubt that this stretch of DNA was acting like a master control gene for sex, so they named it SRY, for *s*ex determining *r*egion of the *Y* chromosome. The cover of the scientific journal *Nature,* in which the discovery was reported, featured a transformed XX mouse displaying its penis.[2] *Science* magazine featured an article entitled "Sex and the Single Gene" with a provocative tag line: "British scientists find all that it takes to make a man is a tiny fragment of DNA."[3]

As popular media predictably followed suit, however, some important scientific details were lost in the shuffle. The discovery of SRY is notable both for what it does and doesn't say about genes and gender. In their original paper, Goodfellow and his colleagues made specific and carefully qualified statements: they asserted that their gene was the only piece of genetic material *found on the Y chromosome* that was required to cause male development in mice. They did *not* say that it was the *only* gene responsible for maleness. "SRY doesn't act in isolation," Goodfellow emphasizes. "It acts by interacting with lots of other genes. Some of those genes regulate the expression of SRY, and some of those genes are regulated by SRY."

The best way to understand this situation is to imagine a chain of dominoes set up on edge in a pattern that divides into two branches. Each domino represents the activity of a single gene involved in sex determination; one branch represents the steps leading toward male development, the other toward female development. The domino representing SRY is located at the branching point.

When the first piece in the chain is knocked over, it begins a reaction; each domino topples the next one in line. If the SRY domino is present, and if it falls in the right direction, the chain reaction proceeds down the branch culminating in the development of testes. If the SRY domino is absent or defective, the action proceeds down the female chain instead.

Similarly, in normal XY embryos once the SRY gene is turned on, it sets in motion a genetic chain reaction that directs the formation of male gonads. When those testes begin to produce testosterone, an entirely separate series of events—which we could represent by another chain of dominoes—is set in motion. If everything proceeds properly in that chain as well, the fetus develops masculine characteristics. In the absence of SRY, on the other hand, ovaries form instead, and in the absence of testicular influence, female characteristics develop.

The domino analogy is useful for two reasons. First, note that "something" must knock the first domino over—it won't fall by itself. Second, in order for the chain reaction to proceed to completion, all the dominoes have to be in line and each one must fall properly. If any "downstream" dominoes are missing or defective, the chain is broken, and the last few dominoes won't fall. Switching back from analogy to reality, David Page and other researchers emphasize that many details of the sex-determining system are still unknown. There is no way to be certain, therefore, that SRY is located *precisely* at the hypothetical branch point. It is also possible that several genes are involved in the testes/ovary decision. For the time being, however, the simplification suggested here is adequate to convey the general nature of genetic control networks.

What has all of this to do with Marilyn Saville? Saville, although an XY female, is not lacking the SRY gene. (She does have testes, remember.) Saville has a defect in another crucial gene, one that affects the body's ability to detect and respond to testosterone. Consequently, she started life just like any other XY embryo. Her SRY gene and the other genetic "dominoes" near the top of the chain were all present, so the first chain reaction proceeded normally. For that reason, she did develop testes, and those testes did produce testosterone. But because of the defective gene she carries elsewhere in her genome, none of the cells in her body could respond to the message carried by that hormone. One of the dominoes in the second chain was missing, so the last links in the chain were never activated.

In consequence, Saville's body acted much like the body of the rabbits whose testes had been removed; the final process of masculinization was halted. No penis formed during development, and her external features are those of a woman. Though the levels of testosterone in her body are those of a normal man, her cells act as though the hormone isn't there. As a result, she looked female, was raised as a woman, and, until her accident, felt herself to be a perfectly normal female.

She is not alone. There are dozens of genes in the chain leading from SRY to maleness. Not surprisingly, there are dozens of different genetic conditions involving defects in those genes that affect thousands of people around the world. Some are sterile but otherwise physically typical men and women. Others are hermaphrodites who carry both male and female sexual organs. Thus, there is a surprising degree of genetic diversity beneath what most of us take for a straightforward either/or situation.

One institution forced to confront this diversity is the International Olympic Committee, a well-meaning group of individuals who try to apply what biologists know about sex to the classifications their events require. In a search for a "simple scientific solution" to what was ultimately a complex human problem, the Committee turned first to genetics and then to molecular biology. What they learned along the way can teach us a great deal about the relationship between genes and the intricacy of the human condition.

The problem began in the 1960s, when Olympic authorities were troubled by rumors that men masquerading as women were participating in athletic events. If such fraud indeed occurred, it would seem to give the perpetrators an unfair athletic advantage because muscles respond to athletic training differently in the presence of testosterone.[4] To make such deception more difficult, female entrants at some competitions were asked to parade nude in front of a panel of doctors. Although many female athletes denounced this degrading practice, others supported the idea of sex testing as a way to protect their female identity from innuendo based on cultural prejudice against female athletes.

Then genetics came to the authorities' rescue, or so they thought. Years earlier, scientists had developed a staining technique that allowed them to remove a few cells from an athlete's cheek and determine whether one or two X chromosomes were present in those cells. Two X chromosomes, it was assumed, would be found only in women. But as you will expect by now, soon after it was introduced, the test turned up several XY women. These conditions are unusual, but not so rare that Olympic authorities can ignore them: one in 500 athletes fails the X chromosome test. Whether the women in question lacked SRY, were insensitive to testosterone, as Marilyn Saville is, or had another, equivalent condition should have been immaterial, at least as far as athletic competition is concerned; although they were sterile, they had normal female anatomies. However, in the eyes of the Olympic Committee they were "men" and as such were barred from competing in women's events.

The International Olympic Committee, aware that its system wasn't perfect and inspired by reports about SRY, decided in 1991 that it had finally found the ultimate sex test. At the 1992 winter games in Albertville, France, they made use of a DNA test for that gene. But prominent Spanish geneticists refused to perform SRY tests at the 1992 summer games in Barcelona, arguing that such tests would misdiagnose a number of legitimate female athletes as men. Other experts in the field, including Goodfellow himself, agree that such a test would only make the problem worse, and not just because of errors in the procedure.

"I'm quite disturbed by people doing sex testing in athletics," Goodfellow declares, "and in particular I don't agree with testing for SRY. I think it's an example of woolly thinking. They haven't defined what it is they're trying to ascertain by using SRY for sex tests. A good example of the problems would be XY women who have mutations in SRY. They would score as positive [male] on this test, but they are women. They have no advantages over any other woman when it comes to athletics."

At least one sports organization, the International Amateur Athletic Federation, agrees. A group of athletes and scientists assembled by the IAAF in 1992 recommended that the only appropriate way to determine sex for athletic purposes is physically to examine the person's external sex organs. Assuming that such examinations can be carried out with a protocol that does not demean or offend participants, it seems the sanest and most sensible course to take. In the meantime, Spanish athlete María José Martínez Patino fought a humiliating three-year battle for the right to compete as a woman. Patino, an XY woman disqualified by genetic sex testing from Olympic competition in the late 1980s, finally won the right to enter the 1992 summer games in Barcelona.

Unfortunately, these developments came decades too late to salvage Marilyn Saville's Olympic dreams, but she has long since come to terms with her life and the people around her. Saville believes that she had to learn about herself and her condition before she could find happiness. Not too long ago she met a man who treats her as a very special, yet ordinary person. Together they share a passion for vintage cars, which they maintain and show at expositions around the country.

To other people in situations like hers, Saville's counsel is simple. "Be frank, be honest, be loving, and treat yourselves right," she advises. "Secrecy just strangles things. It makes you feel very put away and alone. We get along better if we have friends and family who support us and understand. They don't have to pity us, they just have to understand. A little compassion goes a long way."

Saville's emphasis on relationships testifies to the importance of society in the lives of people whose genetic makeup makes them "different" in any way. As any biologist will testify, genes do not operate in isolation within a person's body; they always express their effects in relation to that individual's environment. For humans today, the "environment" is society, not the rigors of life in the wild. Marilyn Saville's genes shaped a healthy body; their action neither makes her ill nor prevents her from enjoying life. In a strictly biological sense, all her genetic condition does is place her in the same situation as the 10 percent of "normal" women who are unable to bear children.

But over and above Saville's biological sex is the issue of her sexuality in a cultural context. It was society that first raised her as a woman and then rejected her on discovering that she did not fit easily into the accepted male-female dichotomy. No gene can be held responsible for the beatings and abuse she endured at the hands of her first husband, or for the rejection she suffered from coaches and teammates.

Marilyn Saville's case carries both a general message about the relationship between genes and society and a specific lesson that holds true for other situations to be explored in later chapters. By itself, better scientific understanding of how genes work may or may not improve the quality of life for people who carry them, depending on how society reacts to their condition. In a more specific arena, this case reminds us that no amount of molecular genetics is likely to succeed in demystifying sex—or any similarly complex aspect of the human condition—a certainty for which we can all be grateful.

• • •

THE STORY OF SRY and sex determination is a powerful demonstration of how a single master control gene can affect a trait as important, and as complicated, as gender. But SRY is just one example of a gene that operates in the upper echelons of the genomic hierarchy. There are many others that regulate gene expression both during embryonic development and throughout life. Furthermore, the information available on SRY to date hasn't revealed much about the real nuts and bolts of its genetic control system. Although researchers are confident that SRY functions as a high-level genetic switch, they still have no idea how SRY itself is regulated or how it turns other genes on or off.

We do have some idea, however, of how genes with more straightforward effects are turned on and off in simpler genetic systems, thanks to the efforts of Nobel laureates François Jacob and Jacques Monod during the 1950s and 1960s. Their search for a direct method to study gene regulation led these men to ground-breaking work in bacteria that laid the foundation for much of modern molecular biology. Why was their work so important? Because the principles of genetic control systems they described apply not only to bacteria, but to multicellular life as well.[5]

Jacob and Monod's work followed from a simple observation: the bacterium *Escherichia coli* can digest milk sugar (lactose) by producing enzymes which break that sugar apart. Those enzymes are produced only in minute amounts by bacteria grown in a culture medium devoid of lactose. Yet whenever *E. coli* is offered that sugar, it begins to produce the enzymes in abundance. To researchers, the question was obvious: What control system turned those genes on and off? The answer was the first conclusive demonstration of a mechanism by which one gene regulates another.

The three genes that enable the digestion of lactose are lined up together on one chromosome. To one side of those genes is a stretch of DNA called a *control region*, because it determines whether or not those genes are turned on. How does this system work? When the message from the control region says "go ahead," the information those genes contain is transcribed onto molecules of messenger RNA, the first step in protein synthesis (see Chapter 1). When the message from the control region, on the other hand, says "stop," no messenger RNA is produced from those genes. As a result, none of the proteins they code for can be made.

Jacob and Monod then found a separate *regulator gene*, which influences the "decision" made by that control region. How does the regulator gene accomplish this feat? It codes for a protein that recognizes and attaches or "binds" to the control region. We now refer to this type of protein as a *DNA-binding protein;* its job is to act like a switch that turns transcription on or off.

In the case of the *E. coli* genes we've been discussing, the regulator gene produces a protein that normally turns the control region off. As a result, the lactose-digesting enzymes are not normally produced in the cell. If there are any lactose molecules floating around, however, the situation changes. Why? Because this particular DNA-binding protein also binds to lactose molecules. Thus, if there are lactose molecules around, the DNA-binding protein attaches to them and is effectively taken out of circulation so it can't attach to the control region. In effect, lactose molecules turn *off* the off switch, and the genes that produce lactose-digesting enzymes are turned on.

This particular regulator gene is fairly low down in the genetic hierarchy of *E. coli;* it appears to control only the genes related to the digestion of lactose. But other genes, higher up in the genetic hierarchy of *E. coli* and other organisms, act as master controls in much the same DNA-binding manner. Some master controls work through the sort of process we've just described; they produce a stop message at the control region, which turns the affected genes off. Others work the opposite way; the DNA-binding proteins those genes produce give control regions the go-ahead signal and therefore turn genes on.

Of course, the situation in humans is more complicated, because we have many more genes than bacteria do. What's more, only a small fraction of our genes are active in any given cell, and they must be turned on in the right place, at the right time, and for the correct length of time. That's why human genes usually have at least five such control regions, and may have as many as twenty rather than just one. In some cases, proteins that attach to these regions may also attach to a sort of molecular gene-activating machine; when the right number of the right kinds of DNA-binding proteins attach to the right control regions, they cause transcription to begin.

After this sort of information about genetic control accumulated, researchers were eager to apply it to influential master controls like SRY. The problem was that at first no one knew what kinds of DNA sequences to look for, either in the vastness of the human genome or in the smaller, but still sizable, genomes of other animals. The breakthrough was made during the early 1980s by researchers working with *Drosophila,* the common fruit flies you might find dancing around on your banana on a summer morning. The background for their work had been laid down over more than half a century by the intellectual heirs of Thomas Hunt Morgan, an American who began studying fruit flies around 1908 at Columbia University. Thanks to his work, *Drosophila* became geneticists' experimental animal of choice for decades.

Over the years, scores of research teams have studied and bred these insects, tracing normal and mutant genes through countless generations. As a result, *Drosophila* is better known genetically than any other multicellular animal. Undistinguished to the naked eye, ordinary fruit flies are transformed under a microscope into living works of art. They bear jewel-like compound eyes, a pair of fragile iridescent wings, and six delicate legs covered with rows of tiny bristles.

The particular flies that led the way to revelations in the study of development, however, were about as uncommon as fruit flies could be. One has legs growing out of the middle of its head, sprouting from the places its antennae should be. Another has an extra set of wings growing out of its body behind the normal pair, crippling the insect so it cannot fly. In the wild, these flies and other mutants like them would quickly die. But in genetics laboratories, pampered and protected like kings and queens, they pass their genetic anomalies from generation to generation.

Why are these mutant flies so valuable? Though it may seem contrary, geneticists who want to study a process in an organism usually search for—or create—individuals in whom the process doesn't work as it normally does. Typically, researchers expose subjects to radiation or chemicals that cause mutations in DNA. They then search among the descendants of these creatures for any whose abnormalities are useful. Geneticists studying cell division look for cells that can't divide correctly; those studying memory look for individuals that can't remember anything; those studying muscle structure look for animals that either can't move or twitch uncontrollably.

Such cases teach researchers how genes normally work, much like automobile breakdowns teach naive drivers how their car's engine works. What's a fan belt? Who knows, until one breaks in hot summer traffic and the engine boils over. Suddenly the driver is not only aware that fan belts exist, but also knows what they look like and how they work. Similarly, geneticists often have no way to functionally dissect a complex genetic system until they find an animal with a "broken" part. That's why these bizarre flies are treasured by geneticists; each bears a mutation that inactivates a single gene in a vital developmental pathway. The result is a whole body part that grows in the wrong place, or a body segment that is duplicated. These genetic variants are called *homeotic* mutations, after the Greek word for "similar," because they generate similar, but misplaced, body parts.

Homeotic mutants were first noticed many years ago, but it wasn't until the late 1970s that Edward Lewis, a prominent *Drosophila* geneticist at the California Institute of Technology, discovered that several genes responsible for these mutants were located in clusters. During the early 1980s, Lewis and other researchers, including William McGinnis of Yale and Matthew Scott of the University of Colorado, succeeded in isolating, sequencing, and manipulating those genes. When they did so, it became clear that these genes shared a distinctive stretch of DNA. These genes came to be called homeotic genes, and their shared sequences were dubbed *homeoboxes*. Their discovery transformed the study of development.

The base sequences within the homeobox, it soon became clear, code for a sequence of sixty amino acids within the larger protein that each homeotic gene produces. Those amino acid sequences excited geneticists for two reasons. First, because homeoboxes seemed to be important in development, researchers searched for those sequences elsewhere in the fly genome. Sure enough, homeoboxes were soon discovered in several other fly genes, most of which are involved in embryonic development. This striking coincidence reinforced the belief that homeoboxes were the Rosetta stones of developmental biology, keys to the mysterious language of development that had puzzled biologists for ages.

Another reason for excitement was that the proteins coded for by homeobox sequences seemed to work in much the same way as parts of the regulatory proteins that Jacob and Monod found in bacteria and which other researchers found in yeasts. To biologists who understand the "grammar" of proteins, the structure of these amino acid sequences suggested the molecular phrase "I will bind to a stretch of DNA."

Why was that important? Jacob and Monod's work in bacteria had provided the first demonstration that one gene can control the action of another by producing a protein that binds to DNA. Although many researchers hoped that discovery would apply to other organisms, there was no way to know that it would. Work with homeotic genes gave solid evidence that genetic control mechanisms in flies—and by implication in other organisms—all worked the same way: by producing DNA-binding proteins that turn genes on or shut them off. Homeobox genes could therefore be master switches, like SRY, that regulate many other genes lower in the control hierarchy.[6]

Researchers' hopes were boosted by the discovery of homeobox-like sequences nearly everywhere they looked for them—in worms, beetles, frogs, chickens, mice, and humans. Even more startling was the realization that those genes weren't strewn randomly about in the genome. Instead, comparable homeotic genes were arranged in remarkably similar clusters in flies, mice, and humans. To everyone's surprise, the arrangement of genes in each cluster related directly to the parts of the embryo in which those genes operate; genes that control head development were found at one end of the cluster, while those that control the tail were at the other.

These extraordinary similarities among organisms that last shared a common ancestor more than 500 million years ago argue that something about these developmental control genes is vital to survival. Apparently the genetic control hierarchy was set up very early on in the evolution of multicellular organisms. But what sorts of decisions was that system making? And how do all these studies in fruit flies relate to humans?

For answers, we turn to the stories of two children, Jeffrey and Zack. Their unusual conditions remind us that among the earliest and most fundamental decisions made as a fertilized egg begins to divide are three so basic that we take them for granted: what begins as a sphere soon develops a head and a rear end, a top and a bottom, and a left and a right side. As you will suspect by now, those decisions are moderated by several series of interacting genes. When that genetic hierarchy slips up, the results are both scientifically fascinating and medically dangerous.

At one year of age, Jeffrey is a stocky, healthy, active, and perpetually curious baby. Unless you saw him during one of his visits to the Vancouver Children's Hospital, you would never guess that he was different from any other child his age. Yet at the hospital, Jeffrey is a celebrity. X rays reveal that he is put together in a dramatically different fashion from the rest of us: his body has mixed up its right and left sides. That doesn't mean simply that Jeffrey is left handed; it means that all his internal organs have switched sides.

That's remarkable, because although the human body looks symmetrical externally,[7] our internal anatomy is dramatically different on the right and left sides. In 99.9 percent of humans, the right lung has one more lobe than the left lung, the liver is located on the right side of the abdomen, the spleen and heart are on the left, and the blood vessels leading to and from the heart on right and left sides are dramatically different. In fact, all mammals and most animals with backbones are asymmetric inside, a trait that seems to date back at least 500 million years.

The process that controls the body's asymmetry, however, doesn't always work as expected. It certainly didn't in Jeffrey's case. "Jeffrey," explains physician George Sandor, "has decided to have his heart on his right side and his liver on his left side. But everything is working well. The blood is going where it ought to go, and there's no problem at all with his heart. Jeffrey, apart from having a mirror-image sort of system, is absolutely normal inside, and he should be absolutely normal forever. As long as the plumbing's fine, we really don't care, and the plumbing's fine." Sandor emphasizes that Jeffrey's condition presents no problems because his left and right sides have switched *completely*. As a result, all his internal organs developed correctly with respect to one another.

Little Zachary, however, is a different story. Blond, pale, and small for his age, he suffers from dextrocardia, a condition in which only his *heart* has confused left and right while the rest of his body developed normally. The result has been disastrous, because the developing human heart is particularly sensitive to its position and orientation in the body. The heart first forms as a tube in the middle of the embryo. That tube bends to the right, doubles up, and then drifts to the left while forming four chambers. Meanwhile, the major blood vessels, which develop separately, meet up with the growing heart and connect with the appropriate chambers. But in cases like Zachary's, where left and right are confused, either the wrong connections or no connections at all are made.

To make matters worse, Zachary's heart is badly deformed: with only two chambers instead of four, it wasn't able to pump blood through his lungs to pick up oxygen when he was born. At birth he required an immediate operation to rectify the problem. After two more operations, doctors hope he will be able to live a "fairly normal" (though decidedly sedentary) life for fifteen or twenty years. Beyond that point, they are unwilling to predict.

The experimental animal that provides us with the most insights into the sorts of genetic decisions that govern these early and fundamental decisions during development is again the fruit fly. A fly embryo, although it develops through a sequence of events strikingly different from human embryos, must solve the same fundamental problems: it must distinguish head and tail, back and front, right and left sides, so that its growing tissues "know" where to put its various body parts. We now know a great deal about those developmental decisions through the elegant experiments of Christiane Nüsslein-Volhard, a geneticist at the Max Planck Institute for Developmental Biology in Tübingen, Germany.

Until Nüsslein-Volhard began her work in 1980, no one had studied the genes that govern these vital early events. That wasn't just an odd oversight; it was an unavoidable result of geneticists' habit of working with adult flies. The problem was that mutations in genes which do their work early during development produce deformities that make four-winged flies look normal by comparison. And if a genetic change destroys a function as fundamental as determining front from back, the unfortunate egg that carries it won't survive to adulthood. As a result, researchers looking only at adult flies found very few mutations with major effects in the first stages of development.

Nüsslein-Volhard, on the other hand, studied eggs, embryos, and larvae to find mutations that would cause gross malformation and early death. Despite the fact that many of her important mutations were lethal, she was able to perform her experiments because these particular alleles were recessive, so flies with a single "good" copy of the gene were normal—and survived to reproduce. Therefore, once she found a gene, she could keep the experimental line going by culturing the living brothers and sisters of the dead flies, animals that carried only one copy of the mutated gene.

As Nüsslein-Volhard recovered dead embryos and larvae, she determined at which stage development had gone awry, that is, at which point the dominolike genetic sequence of events had been halted. By studying one such mutant after another, she and her colleagues assembled an intricate picture of the chain of events that transform a fertilized egg into a complex, segmented fly embryo. One mutant, for example, dies because it has two tail ends and no front; the mutated gene seems to help specify the front end of the animal. Another mutant is missing half its body segments, and a third is missing the middle of its body.

As it turns out, the egg knows front from back even before it is fertilized. How? The cells of the mother fly that produce and nurture the egg insert the RNA product of one gene into one end of the elongated egg cell. That RNA remains near the end where it entered, but the genetic message it carries is converted into a protein that slowly diffuses toward the opposite end of the egg. In the meantime, RNA from a different maternal gene moves in the other direction. Each of these gene products turns certain genes on and other genes off when the egg begins to grow. Thus, because these master control compounds are present in different concentrations in various places, different genes turn on and off in various parts of the embryo.

That process starts the dominoes falling and the developmental patterns branching. The first genes to be activated turn on other genes. Those genes activate another set, and so on down the line. The end result is an embryo with eighteen separate segments, each of which relies on different genes to produce appropriate body parts. Head segments produce mouth parts and antennae, thorax segments produce wings and legs, and abdominal segments carry organs such as gonads.

This, of course, is just part of the story. Developmental signals of several kinds are constantly passing through different parts of the embryo, setting up a cascade of activity in which timing is vital. You may now be able to imagine how an inappropriate response to a signal factor, or a factor present in the wrong place or at the wrong time, can cause errors of the type that place legs on top of the head.

Human embryos are much more complicated, and we therefore have only a rudimentary understanding of how the equivalent process works in our species. Some reasons for that lack of knowledge relate to the problem that delayed work with flies until Nüsslein-Volhard came along. What would happen to a human fetus if it confused head and tail? We would never see such unfortunate individuals, because they would die before birth. And of course, ethical considerations, thankfully, forbid us from treating human eggs the way Nüsslein-Volhard handles flies.

Certain situations, however, offer some tantalizing clues about early events in human embryos. As the cases of Jeffrey and Zachary indicate, confusion of left and right side can sometimes produce viable babies and at other times produce medical problems that interest both clinicians and basic researchers. In a fascinating parallel development, geneticists have found a mutation that produces apparently similar left/right confusion in mice. Half the mice that carry two copies of this gene are born with their left and right sides reversed, as Jeffrey's are, while the other half are normal. In addition, a significant number of newborn mice with that genetic makeup suffer from severe heart defects akin to Zachary's.

Researchers are close to locating this gene in mice and hope that once they do, they will be able to find an equivalent stretch of DNA in humans. They hypothesize that an individual who inherits two copies of this mutant allele cannot follow the normal instructions that distinguish left from right in the early embryo. It still isn't clear just how the gene might work, but there are further clues from the study of human Siamese and identical twins, who emerge when a single fertilized egg splits to produce two individuals with identical genes. Identical twins separate early enough in development that both are often perfectly healthy, although they may share the same placenta and umbilical cord in the womb. Siamese twins, on the other hand, separate late enough that they remain joined by some part of their body even after birth.

It has long been known that some Siamese twins are literally mirror images of one another; one develops normally, while the other is right/left reversed as Jeffrey is. It is also known that if Siamese twins are joined side by side, the left-hand twin is usually normal while the right-hand twin may have heart problems. Even more intriguing is the fact that some identical twins, although their internal anatomies are normal, carry what researchers call bookend traits; they suck opposite thumbs, develop teeth in alternate ways, and grow hair that spirals in opposite directions on top of their heads.

Remarkably, one British researcher who studied twelve pairs of twins with heart irregularities found eight sets of identical twins in whom problems all occurred in the twin with the hair spiral on the right side of the head. This leads some scientists to suggest that the system which determines orientation in mammalian embryos may be analogous to that described in flies. It seems as though some sort of compound may be present in the embryo in greater concentration on the left side than on the right, offering clues to proper position.

How might this system work in twins that form from left and right halves of the embryo? The left-hand twin, who develops from the area that carries the highest concentration of this hypothetical compound, receives the proper orientation signal and develops normally. The other twin, separated from the source of the compound by the gap between twins, doesn't receive the proper signal and develops either left/right confusion or heart problems. If this hypothesis is true, the gene that researchers are seeking in mice may be the "switch" that responds to this compound. It is clear in any case that this gene would be just one link in a longer chain, just as SRY is only one piece of a complex genetic sex-determination system.

Still, the notion that control of development might be so similar in humans and fruit flies, while exciting and intuitively obvious to some researchers, seems extremely unlikely to others. The gap between *Drosophila* and *Homo sapiens*, critics argue, is simply too large for that to be true; we develop in so many different ways into such different creatures.

In an effort to bridge the gap between insect and human, Nüsslein-Volhard and her colleagues are adding racks of aquaria to their collections of fruit-fly bottles. Those aquaria are filled with zebra fish, small, lively animals that share more recent common ancestors with us than either insects or nematode worms. Fishes, like humans, birds, and reptiles, are vertebrates, and researchers know from more than a century of embryological observations that many visible stages of their early development resemble ours more closely than those of any other convenient genetic experimental model except mice. What's more, zebra fish eggs and larvae are transparent, so although their bodies contain far more cells than nematodes, many steps in their development are easy to observe. With luck, zebra fish studies will soon disclose either similarities or differences between our developmental genes and those of the well-studied flies. In either case, they are bound to help us understand at least some of the first stages in early development.

These studies are of more than just intellectual interest. Although understanding genetic mechanisms of development probably won't produce cures for children like Jeffrey and Zachary, and may or may not lead to treatments for people born with Saville's condition, they are already providing invaluable information related to other medical problems. Studies of development concentrate on the constant and dynamic communication among cells, during which various molecular messengers turn regulatory genes on and off. As we learn more about those messengers, which range from human growth hormone to a much simpler compound called retinoic acid, we are finding that it may be possible to prevent or moderate certain effects of biological senescence, from the loss of muscle mass to the aging of skin. If we could only understand how to "reactivate" certain genes normally turned off during development, we might be able to encourage the stumps of severed human limbs to repair themselves and replace their missing parts, as those of some animals do. And if we can improve our understanding of the chemical pathways that regulate cell growth, we may be able to conquer one of humanity's oldest and most intractable ailments. We have a name for cells that ignore the body's normal commands to differentiate and stop reproducing; we call them cancers.

4

Conquering Cancer

ON A WARM, breezy day in California's Napa Valley, a crowd gathers on a shady lawn to celebrate Alfredo and Giulia Santi's fifty-fourth wedding anniversary. Two of their great-grandchildren, Dominic and Brianna, scamper around in their party clothes. The rest of the guests eventually settle down at long tables set up beneath an open tent. An elaborately decorated cake arrives, inscribed with lilac letters: "Happy Anniversary! 54 years of stirring the pot!" A portrait of Nonna and Nonno, Italian for Grandma and Grandpa, adorns the center. Alfredo Santi points at his likeness. "Lookit," he complains to four-year-old Brianna. "No hair at all!" She erupts with laughter.

These people are clearly a family, a close-knit one at that. Kinship is obvious in the affection lavished on children, the intimate yet animated conversations, and the respect accorded to the patriarch and matriarch at the head table. Strong physical resemblances unite them, too, in visible confirmation of the fact that Alfredo and Giulia Santi's genes live on in their descendants.

But along with facial features and smiles common to both sides of the Santi family comes another, less felicitous legacy, breast cancer. That dreaded disease has already struck three women at the gathering. Katherine, a vibrant, raven-haired woman, was diagnosed with breast cancer at the extraordinarily early age of thirty one. Her mother, Iris, contracted it at thirty-nine. Iris's sister Julie was stricken at forty-two. And they were not the first in the family to suffer the effects of this scourge; cancer also claimed the lives of Alfredo's mother and a sister. Looking at young Brianna, Katherine's daughter, it is hard not to wonder about her future as well.

THE COLD NORTHEAST wind ruffles Connie McNally's short brown hair as she leaves her oncologist's office at Massachusetts General Hospital in Boston. Like the women in the Santi family, McNally is battling breast cancer, but her disease, unlike theirs, does not run in her family; physicians have told her that she need not worry about passing it on to her daughter.

That consolation notwithstanding, Connie McNally's condition is serious, but she is neither ready to give up nor about to withdraw into herself. To help others in her situation, she works at a support center for cancer patients. To help herself, she keeps up with advances in cancer research that might improve her chances of recovery. To some extent because of that effort, she is about to participate in the first clinical trial of a new therapy that may revolutionize treatment of certain types of breast cancer. In the meantime, her personal battle against cancer is a race against time. Will she live to see her daughter's next birthday party? Her own wedding anniversary? Each event she reaches is a victory of sorts, a small concession wrung from the disease.

McNally and the women of the Santi family are not alone in their fears, their courage, or their race against time. Cancer in one form or another claims 1,400 lives *each day* in the United States alone. Most of us know someone who has either conquered cancer or succumbed to it. Some of us are contending with it ourselves. Others of us harbor tiny, potentially dangerous tumors within our bodies, but aren't yet aware of them. Cancer spares no race, class, or age group. Retinal cancer strikes children. Breast cancer, prostate cancer, lung cancer, colon cancer, skin cancer, and many other cancers attack adults.

We have heard about "progress in the war on cancer" ever since that battle was declared in 1971. But according to many news reports, certain forms of cancer are on the increase; invasive breast cancer incidence seems to have risen dramatically from 1980 to 1987, reaching what many women's health experts have called epidemic proportions. One current National Cancer Institute brochure proclaims "Good News: Everyone does not get cancer. 2 out of 3 Americans never will get it."

But there is encouraging news behind the grim headlines and twenty-second sound bites. Many cancer epidemiologists assert that apparent increases in breast cancer incidence are the result of greater vigilance among women and increased use of mammography, which together have led to the earlier detection of many small tumors, and some benign ones, that previously went unnoticed for years. The incidence of numerous other cancers, with the notable exception of those induced by cigarette smoke, has been almost constant. However, clinicians note that despite the increase in reported cases of breast cancer, *mortality* from that disease has been fairly flat, while the cure rate among some cancers, including colon cancer, is slowly rising.

Still, why is it taking so long to defeat this many-faced threat? Has any *real* progress been made? What does our current knowledge mean to someone who has just been diagnosed with a malignant tumor? For that matter, what exactly *is* cancer?

Historically, cancer has been difficult to understand for two main reasons. First, the generic term "cancer" encompasses dozens of different tumor types, which occur in different tissues, behave in slightly different ways as they grow, and respond differently to surgery and chemotherapy. Second, these cancers seem to be provoked by a host of apparently unrelated causes. Some appear to be triggered by exposure to sunlight or other forms of radiation. Ongoing studies suggest that 30 to 50 percent of cancers in the United States can be correlated with diet, about 35 percent with smoking, and 5 percent or less with environmental carcinogens. But the results of these studies, with the exception of those related to smoking, are not complete, and few have yielded specific recommendations. A small number of human cancers, including certain types of cervical and liver cancer, can be caused by infection with certain viruses, while some breast and ovarian cancers may be tied to childbearing habits. Other forms seem to be inherited, as with breast cancer in the Santi family.

For years this diversity in cause and effect led cancer researchers down dozens of parallel paths and confounded their search for common origins and cures. But the revolution in molecular biology has made possible a major conceptual breakthrough: the realization that in cancer, as in life itself, there is unity behind the diversity. Researchers now believe that all cancers can be traced back to damage in DNA that causes cells to disregard signals that normally tell them how to behave as proper citizens within the community of the body. Once those controls are ignored, cancer cells become pernicious free agents, parasites that reproduce out of control, spreading through the body and ultimately killing their host. "We contain within us the seeds of our own destruction," Michael Bishop of the University of California at San Francisco informs us solemnly. "Cancer is within our genes."

Some cancers, like those which appear so frequently in the Santi family, seem to be triggered by a defective gene that can be passed from generation to generation. Researchers believe that Alfredo Santi inherited such a gene from his mother, although they haven't yet identified it. It poses little direct threat to Alfredo, because breast cancer rarely strikes men; according to the National Cancer Institute, 300 men died of breast cancer in 1991, compared with 44,500 women. Nonetheless, every cell in Alfredo's body, including those which produce his sperm, carries that gene. That's how he unwittingly passed it on to two of his three daughters. To those women, the gene poses a deadly threat; researchers suspect that it predisposes them to develop both breast cancer and ovarian cancer at an early age.

The vast majority of cancers, however, although they are also caused by defective genes, are not inherited. Connie McNally, for instance, seems to have inherited a good set of genes, but at some point during her lifetime, critical genes in one or another tissue in her body were somehow damaged. The progeny of the original damaged cell or cells are the only cells in her body that carry the rogue DNA whose actions threaten her life. Because the egg cells in her ovaries are free of it, she cannot bequeath it to her children.

Working from that unifying genetic understanding, researchers today are identifying, isolating, studying, and manipulating genes related to cancer in their efforts to cut off the disease as close to its roots as possible. The challenge is staggering; it requires unprecedented insights into the workings of genes, equally sophisticated knowledge of a maze of molecular machines that control cell growth and division, and the courage to experiment with some of life's most fundamental processes in living human patients.

Despite that difficulty, work in understanding the causes and origins of cancer is progressing more rapidly than ever. Scores of researchers working with experimental organisms, ranging from yeast to mice, are taking advantage of life's molecular unity to study the basic processes that tell cells when to grow and when not to. Other studies focus on the body's innate mechanisms for fighting rebellious cells, aiming to understand the system well enough to boost its efficiency. Piece by piece, findings from those apparently disparate lines of research are converging and adding up to a tantalizing series of clues. "Most of us," asserts University of Michigan geneticist Francis Collins, "think we are perched on a threshold. People will look back on this era one hundred years from now and say, 'Boy, that was the time to be working on cancer! It was all solved twenty years later, but that was a great time to work on it as far as the joy of discovery and the excitement of advances.' "

Marc Lippman, a researcher at Georgetown University School of Medicine, agrees. "The knowledge we have acquired in the past six or seven years," he declares, "triples *all* previous knowledge. It's my personal belief that over the next decade, new treatments will radically change our approach to the disease. Thousands of people will have much better therapies that will alleviate their suffering and, in many cases, cure the disease. I couldn't be more excited."

But just what do researchers hope to learn by studying the Santi clan and others who pass cancer from one generation to the next? How does that knowledge relate to people like Connie McNally, whose cancers are not inherited? And how quickly will any of this knowledge actually translate into treatments and cures? Answering those questions requires a molecular understanding of cell growth and development, processes which reveal the actions of the vital, yet potentially deadly, genes at the root of cancer.

• • •

EARLY IN EMBRYONIC LIFE, cellular activity progresses with remarkable speed: after the fertilized egg begins to divide, cells migrate from one spot to another, form large masses, or spread out into thin sheets. Some cells build muscles, others form nerves, still others produce skin. This riot of activity is choreographed by a host of growth enhancers and growth inhibitors, all of which are produced at the behest of genes, and all of which interact with other genes that turn on and off in different parts of the growing fetus.

Later on, this hustle and bustle slows down, as many cells cease moving around, assume mature shapes, and stop growing. They do so not because they somehow run out of steam, but because certain genes direct them to stop dividing and moving, while other genes instruct them to assume their mature shapes and functions. Those genes are influenced in that decision by a complex series of interacting messages which somehow enable the cell to sense its context in the body.

But some cells keep growing and dividing throughout life, including one large and important class of cells that form both the skin that covers our bodies and the lining of many internal organs, including the colon, uterus, lungs, and milk ducts in the breast. These are called epithelial cells, and the tissues they make are called either epithelium or epidermis. The fact that some of them keep growing throughout life is of special interest, because from them spring the vast majority of human cancers—up to 96 percent, according to some estimates. Several other cell types keep growing during adult life; these, too, are candidates to become cancerous.

Two prominent examples are stem cells in the bone marrow that churn out blood cells throughout life and liver cells.

Normal growth in all these organs is predictable. Stem cells living deep within the tissue somehow retain many properties of embryonic cells throughout life, including the ability to divide and form two daughter cells. One of these daughters, which remains where the mother cell was, can become a mother cell itself by dividing at some time in the future. The other daughter cell, which "leaves home," is gradually pushed upward by other, newer cells produced beneath it. As that second daughter cell matures, it assumes an adult shape and performs duties appropriate to its role in the tissue. Skin cells, for example, form a tight barrier, while those of the gut begin to absorb nutrients. Sooner or later, these "floating" cells reach the top of the tissue layer, perform their function for some days, then die, to be shed from the surface of the skin into the intestine or into the milk, depending on where they are.

Normally, the system is that predictable. Stem cells on the bottom keep dividing at a rate that fairly evenly replaces loss of mature cells on the top. Those upper cells never divide; somehow, their copies of the genes that enable cell division are turned off, in most cases irrevocably. Those genes and their instructions are still present, but their information is locked away by the chemical messages that circulate around them.

But if the tissue is damaged—if, for example, a razor blade slices into your finger—a gaping chasm suddenly separates skin cells that were once neighbors. Because of the empty space beside them and because of other body responses to injury, the messages received by stem cells next to the wound change dramatically. Soon critical genes are turned on and once-placid cells leap into action, dividing more rapidly and moving to fill in the gap. There's a lot of dividing and moving to do; from a cellular perspective, filling a wound with new cells is roughly equivalent to filling the Grand Canyon with apples.

What's more, in order for all those new cells to survive in this "canyon," new blood vessels must grow in to supply them. To make those blood vessels grow, certain genes begin producing messenger proteins that stimulate the growth of new capillaries into the wound. Ultimately, when the wound heals, the body's chemical messengers register that fact. Genes that encourage division, movement, and blood vessel growth are turned off, and the tissue returns to normal.

In cancer, however, something peculiar and deadly occurs: somehow, damage to a series of genes causes cells to behave as if they were always near a wound. The genes that enable these cells to divide, move, and encourage the growth of new blood vessels remain constantly active, inciting the cells to proliferate out of control—as if to fill a gap that does not in fact exist. The very same sorts of genetic controls that enable the body to heal itself become, as Bishop described them, "the seeds of our own destruction."[1]

Cancer usually begins when one of those stem cells, whose DNA has been damaged in a critical way, somehow shrugs off normal growth constraints and begins to divide. Then both of *its* daughter cells divide. And *their* daughters keep dividing. Each time these cells divide, they double their numbers. Typically, they also fail to assume their proper shape and refuse to perform their normal functions. Not only is their growth inappropriate to the tissue, but they begin to compete with normal cells for the body's resources. Soon a tiny lump appears: a tumor has been born, but it is not yet dangerous. In fact, it may never cause harm. It has taken only the first step toward malignancy; if nothing else happens to its DNA, it may never become life threatening. Roughly 50 million Americans are walking around with this type of tumor in their bodies, although many of them are unaware of it. Many women have such tiny lumps in their breasts. Nearly all men over the age of seventy have them in their prostate glands. Such tumors, clinically referred to as *carcinoma in situ*, may stay just as they are for ten to twenty years or more.

One constraint on such tumors arises from a simple reality of cellular life: masses of cells can't grow much larger than a BB pellet unless they acquire their own supply of food and oxygen. To do so, at least some cells in a tumor must switch on genes whose products encourage new blood vessels to grow around them and into the small tumor mass. Once that switch is activated—supplying the tumor with blood—the cancerous mass can grow more rapidly, doubling in size within a few months.

Still, if tumors at this stage acquired no further genetic changes, many would be manageable. Treatment for a fast-growing but self-contained tumor would not be pleasant or risk-free by any means, but as long as the collection of rogue cells hasn't compromised the function of a vital tissue, it could be dealt with. Surgical removal of a body part is never trivial, but at least surgery would end the problem; if the entire tumor could be removed, it would present no further threat to life.

But according to some researchers, by the time tumor cells acquire their own blood supply, they have begun to acquire mutations with unusual speed. It seems that the "proofreading" and "repair" mechanisms that normally protect the integrity of DNA don't function well in tumor cells. As a result, mistakes that are made when DNA is copied aren't corrected, and mutations accumulate rapidly. A high mutation rate is dangerous, for sooner or later at least some tumor cells acquire changes in their genes that permit them to migrate much like cells in an embryo—except that these rogue cells pay no attention to the body's control signals and ignore internal geographical boundaries. Ultimately, some of them enter the circulatory system, evade the surveillance of the immune system, and actively invade other parts of the body.

It is cancer cells' ability to migrate, survive, and invade other body tissues that makes them so lethal. Normal cells simply cannot act that way. A fetus in the womb of a pregnant woman, for example, sheds many millions of normal cells into its mother's blood during the course of development, but those free-floating cells never cause trouble because they simply do not survive. But malignant tumor cells do. And because they survive, they spread throughout the body, planting new tumors in so many places that doctors can't find and destroy them all. This insidious progression and spread of tumor cells—*metastasis*—ultimately kills 20 percent of cancer patients.

What are the genetic controls behind renegade cell growth and metastasis? What interferes with their normal activity? Can they be returned to normal? Are there any other ways to fight the problems they cause? Answers to these questions and the techniques used to find them are as diverse as the faces of cancer itself. Some research teams concentrate on inherited cancers in humans, relying on the help of families like the Santis to help them track down wayward genes. Other teams concentrate on noninherited cancers like Connie McNally's, depending not only on human patients, but on a surprising collection of nonhuman players.[2] If, for example, McNally is lucky, her life just might be saved by a series of experiments performed years in the past on laboratory rats.

• • •

DESPITE HER COURAGE and determination, McNally is in a precarious position. Although her original breast tumors were removed, the cancer had already spread to other parts of her body, most noticeably a region in her chest. Unfortunately, as is often the case with metastatic cancer, aggressive treatment with chemotherapy has not kept those scattered tumors in check.

That's why she made the trip from Boston to Los Angeles, where she is participating in a cutting-edge clinical trial. Although nervous, she is very positive about her role in the procedure. "I'm really excited," McNally declares, "because just two or three weeks back my only options were all chemotherapy related. I have already had a lot of that and I wasn't anxious to put myself through it again. Then I heard about this, and it was completely new and different. So a couple of weeks later here I am! It is all going to happen and there is really a reason for hope. I get to take advantage of the most current research to help me get rid of this. I also feel that it is going to help a lot of other people, which is a double benefit."

The treatment McNally is about to undergo has been pioneered by Dennis Slamon, a physician-researcher at the University of California at Los Angeles Medical School who seems to be in perpetual motion. If he isn't responding to a critical clinical call, he is tending to a research problem. Yet despite his hectic schedule, Slamon manages to appear at once warm, confident, and professional; during the time he spends with McNally, she receives his undivided attention. Slamon speaks forcefully of the helplessness he has always felt when faced with cancers that don't respond to treatment. "As a physician you are supposed to *do* something," he confesses. "And you are presented with these diseases for which you can do nothing with your current armamentarium. It is very frustrating."

It's not that existing treatments are totally ineffectual; chemotherapy can cure certain childhood lymphomas, has a reasonable success rate with Hodgkin's disease and testicular cancer, and is believed by some researchers to extend life span in breast cancer patients significantly. Still, both chemotherapy and radiation therapy, the major weapons against cancer, are less effective against certain types of tumors and can be highly toxic to patients. "These treatments are not really based on any rational understanding of the disease," Slamon points out. "They are tantamount to throwing in a hand grenade and hoping that you kill more bad cells than good ones. That is about as sophisticated as it gets."

Unfortunately, Slamon's analogy is accurate. Chemotherapeutic drugs work by interfering with the duplication of DNA. Because continually dividing cancer cells duplicate their DNA frequently, these drugs are especially toxic to them. Normal cells in the body that are not dividing are hardly affected. But any cells in the body that *are* actively dividing—including the intestinal stem cells we described earlier, cells that produce hair, and those in the bone marrow that produce new blood cells—are also affected and die in large numbers. That's why nausea, hair loss, and blood problems are common and often life-threatening side effects of chemotherapy.

Radiation therapy is not much better. "The way we deliver radiation therapy," Slamon explains, "is to identify a site in the body that we want to irradiate and we aim a beam at it. Let's say you had a mass in a lung; a beam would be aimed at that mass. The tumor may be deep inside. All of the tissue in the way, the tissue behind it, and all the tissue around it gets damaged. More important, there may be groups of tumor cells that aren't large enough to be seen on an X ray but are still there, and you would never know to direct a beam there."

His frustration with these brute-force techniques inspired Slamon to study cancer on the molecular level. His goal, like that of many other researchers, was to use an understanding of the genes and proteins involved in cancer to develop medical "smart bombs" that would bypass normal cells and target cancer cells precisely. Unfortunately, that job is difficult for two main reasons. First, cancer cells are not all that different from other body cells, so it isn't easy to find a feature that sets them apart as targets for therapy. Tracking down the genes that cause cancer is also hard because so many genes have been implicated. "There are different roads to Rome," Slamon explains. "There are different ways to get from a normal breast cell to a cancerous cell, and there are different genes that are altered in different patients."

Still, he knew that even if a gene he found was altered in just a few patients, such a discovery might lead the way toward better therapies for at least that particular group of people. As things turned out, Slamon—and by association Connie McNally and thousands of other women that share her form of breast cancer—were lucky. For Slamon's research was facilitated by a series of experiments that were performed in a laboratory just across the Charles River from McNally's oncologist's Boston office. To understand the experimental treatment that brought McNally into Slamon's research program, we must backtrack briefly to that laboratory and to a line of basic research that may seem, to the uninitiated, to have little to do with breast cancer.

The complexity of research in molecular biology makes it unfair to credit a single laboratory with exclusive credit for any discovery. All researchers build on the work of their predecessors, and at least half a dozen laboratories either collaborate or compete with one another in nearly every area of cancer research. Yet certain teams have repeatedly played pivotal roles in the search for a genetic understanding of cancer. The laboratory led by Robert Weinberg at the Whitehead Institute for Biomedical Research in Cambridge, Massachusetts, is one such research group.

Weinberg is a researcher of whom colleagues speak very highly, partly because of his accomplishments and partly because of his personality. Weinberg holds a Ph.D., rather than a medical degree, and although his work often has clinical applications, the cells he studies frequently come from nonhuman subjects. Weinberg regards childhood hours spent fiddling with electric trains and tinkering with appliances as prologue to his current passion for research. He maintains that he is motivated as much by his curiosity about the way cells work as he is by the challenge of curing cancer. "I happen to be interested in the internal circuitry of the cell," he asserts. "That is my obsession and that of everybody who works with me." Luckily for people like McNally, Weinberg's obsession with genes that control cell division has provided several insights vital to understanding cancer.

Some of those insights can be summed up by comparing a dividing cell with an automobile in motion and likening the rate of cell division to the speed of the car. A cell in normal tissue acts like a car traveling judiciously through a school zone; both proceed quite slowly. The cells in an embryo or near a fresh wound, on the other hand, are more like cars on an interstate; both zoom along at high speed. Problems arise only when either automobile or cell disregards appropriate rules by driving at freeway speeds in a school zone or by dividing at a rate appropriate to wound healing in normal tissue. It takes only one bad driver to cause an accident, and it may require but a single lawless cell to create a cancer that threatens the entire body.

The analogy goes further. The speed of an automobile is controlled by its accelerator and its brakes; both are needed to adjust for changing conditions. Cell division is controlled analogously by two sets of molecular pathways whose actions complement each other. One type of pathway works like stepping on the gas pedal—it stimulates cells to divide. The other is the cellular equivalent of the braking system—it works to slow down or stop cell division. Naturally, both sets of cellular systems are composed of interacting molecular machines, proteins that are built and controlled by genes.

In cancer, mutations in those genes cause them to produce proteins that function abnormally. When the cell's "accelerator" system works properly, for example, it responds to growth-promoting messages by inducing the cell to divide. If the genes that control this system are damaged, they can cause the "gas pedal" to stay depressed by itself, forcing the cell to act as though a growth-promoting substance is present even when it isn't. Malfunctioning genes of this type, called *oncogenes*, are believed to stimulate the undisciplined cell division that lies at the root of many human cancers.

Normal genes that control the "brakes" in our analogy are called *tumor suppressor genes*. When working properly, the products of these genes prevent cells from dividing when they detect the presence of growth-inhibiting messages. Mutations in these genes can cause a cell to ignore a growth-inhibiting signal, with disastrous results. The cell may still receive a message to stop—the molecular equivalent of a red light— but it lacks the wherewithal (functioning brakes) to respond properly. If you've ever been in that unfortunate situation in a car, you can imagine what happens with a comparable malfunction in the cell; once the machinery of cell division is started, there is no way to make it stop.

Many researchers believe that cells must acquire defects in both these systems—in addition to other mutations necessary for metastasis—in order to pose significant threats to health. As you might expect by now, cells are much more complicated than cars; instead of single gas and brake pedals, there seem to be multiple versions of each. Cells have several interconnected molecular systems that stimulate division, and a malfunction in any of them may cause trouble. The same seems to be true of cellular brakes. Because each of these systems is controlled by several genes, the total number of potential oncogenes and tumor suppressor genes is very large, a fact that complicates searches for diagnostic tools and treatments.

Weinberg's laboratory has contributed extensively to the study of both oncogenes and tumor suppressor genes. Among its other accomplishments, his group was one of the first to positively identify, isolate, and sequence an oncogene in mammals. In an important series of experiments, Weinberg studied nerve cell tumors in rats that other researchers had induced by exposing the animals to carcinogenic chemicals. He then ground up the tumor cells to extract their DNA, which he searched for cancer-causing genes. He found one gene, later named *Her-2/neu,* that seemed to be associated only with tumor cells. He then proved that this was, in fact, an oncogene by showing that he could pluck this gene from a cancer cell, place it inside a normal cell, and thereby cause the previously normal cell to grow like a cancer.

But *how* did this gene force a cell to grow malignantly? To answer that question, Weinberg's coworkers read the information coded in the gene and translated that information into the sequence of amino acids which form its protein product. By comparing their information with results of other research groups', they concluded that this oncogene protein was probably what molecular biologists call a receptor, an antennalike molecular machine that sits on the cell surface and receives signals from the environment. When this antenna senses the presence of a growth-promoting factor near the cell, it produces a signal inside the cell that initiates cell division. If we return for a moment to our imaginary molecular language, we might say that the "sentence" spelled out by the normal version of this protein is "I will sit in the cell membrane and stimulate cell division whenever I encounter a growth factor outside the cell."

Ultimately, Weinberg's associates determined that their oncogene directed the assembly of a slightly abnormal form of this receptor. Here, as in the case of Hemoglobin-S, a major functional change in a vital protein is caused by an extraordinarily small change in the spelling of its molecular sentence: the mutant receptor protein differs from the normal form in only one of its 1,600 amino acid building blocks. Yet this structural flaw causes a change in the protein's function that is analogous to an accelerator pedal stuck permanently to the floor. In terms of molecular language, this receptor protein now behaves as though its message says, "I will sit in the cell membrane and tell the cell to divide *whether or not* I encounter a growth factor outside the cell." As a result, the cell's internal machinery receives an unrelenting stream of signals telling it to reproduce.

The story of this particular oncogene might have ended here if the gene was, in fact, found only in the nerve-cell tumors of rats that had been fed a particular carcinogenic chemical. But the molecular unity of life assured that Weinberg's discovery was to be just the first episode in the saga of Her-2/neu. The man who made the connection between Her-2/neu and breast cancer was Dennis Slamon. When he set out to identify a gene involved in breast cancer, he followed a hunch that an oncogene previously identified by other researchers in other kinds of tumors might be involved in his patients as well. To test his idea, he went to the UCLA Medical Center tumor bank, where an enormous freezer stores thousands of breast tumor samples that are saved after surgical removal during mastectomies and lumpectomies. Slamon ground up these tumors, extracted their DNA, and searched among them for mutations in DNA sequences that matched any known oncogenes. By great fortune he discovered that in 20 to 30 percent of the tumors he screened, something strange was going on with Her-2/neu, the very same gene that Weinberg had implicated in rat nerve-cell cancers.

Curiously, Slamon's breast cancer patients didn't carry precisely the same mutation in the Her-2/neu that Weinberg had found in his rat tumors. Instead of a subtle alteration in the gene's DNA sequence, cells from these breast tumors carried extra copies of the gene. Normal cells carry two copies of Her-2/neu, one on each of two chromosomes. But some tumor cells from breast cancer patients had between two and fifty additional copies. That discovery was surprising at the time. Now, however, we know that tumor cell DNA is prone to several types of errors in replication, including gene duplication. For reasons that are still not completely understood, those extra copies of Her-2/neu seemed to produce an effect equivalent to that caused by the "stuck accelerator" mutation; the tumors of patients whose cells carried those extra copies grew more aggressively, were more likely to recur, and were less likely to respond to conventional therapies than those which did not.

That's where Connie McNally first entered the picture. Her oncologist sent Slamon samples of the tumors removed from her breasts. Tests on DNA extracted from those samples showed that her tumor cells indeed carried many extra copies of the Her-2/neu gene. Until recently, that would have been unequivocally bad news for McNally—close to a death sentence, in fact—for Slamon's diagnosis was powerful evidence that her particular form of breast cancer was one of the most virulent. That is how knowledge accumulates in this field; ability to recognize a problem necessarily precedes the ability to do anything about it. To Slamon's dismay, the first generation of patients in whom this genetic anomaly was diagnosed has begun to die.

But for Connie McNally the timing may be better; she may be *just* in time to receive the first real benefits from the discovery of this gene, since Slamon has taken several steps beyond spotting this oncogene in patients. Thanks to the work of numerous research teams, including Weinberg's, Slamon not only knows what the gene's protein product is, he also knows that it happens to be a cell surface receptor. That knowledge is power: it puts Slamon in a position to develop therapies which target that protein. Instead of resorting to chemotherapeutic "hand grenades," he hopes to develop "magic bullets" that kill only cancer cells, just as antibiotics kill only bacterial cells while leaving the body's other cells unharmed.

What might his strategy be? Slamon knows that extra copies of the Her-2/neu gene cause McNally's tumor cells to make many more receptors—as many as 10,000 times more—than normal cells make. He also knows that the extra receptors somehow encourage those cells to divide. But the fact that this renegade protein sits on the cell surface makes it vulnerable. Because parts of all the extra molecules are exposed on the outer surface of the cells that carry them, those particular cells "look" different at the molecular level. If Slamon could only find or construct a molecule that could recognize that unusual concentration of antennae on tumor cells, he might be able to use it either to prevent the receptor molecules from sending their life-threatening message or to single out and destroy only those cells which carry them.

Where could Slamon find a molecular "bloodhound" that could attach exclusively to this protein? Researchers' understanding of protein structure and function was—and is—too rudimentary for him to have known how to assemble such a molecular machine from scratch. Luckily, as we will explain in the next chapter, the immune system just happens to specialize in making molecules designed to track down other molecules. Whenever a foreign protein is injected into the body, the immune system produces molecules called antibodies that recognize its shape and attach to it. Extraordinary skill in manipulating antibodies and the genes that produce them enabled Slamon and his colleagues to produce antibodies custom-tailored to find and attack Her-2/neu receptors.

The first problem they faced was a straightforward one: because the Her-2/neu protein is a normal component of human cells, cancer patients' bodies do not ordinarily produce antibodies against it. Furthermore, the procedures necessary to generate and recover antibodies and the genes that make them involve procedures that cannot be performed on human subjects. For that reason, Slamon began by isolating the human version of the Her-2/neu protein and injecting it into mice. The mice obligingly produced an antibody that recognizes and grabs on to the human tumor cell protein. Then, in collaboration with Genentech, a biotechnology company, Slamon's team isolated the genes that direct the production of those antibodies. The researchers inserted those genes into a culture of mouse cells set up as a biological production line. Those cells, following instructions in the transplanted gene, produced large quantities of the antibody.

Excitement in the lab mounted when an experiment proved that the mouse antibody slowed the growth of cancerous cells in a petri dish. Further experiments showed that it could also stop the growth of similar human tumors which had been transplanted into mice. The next step was clear: a clinical trial in humans. After a long discussion about the risks and benefits of the new procedure, Connie McNally and several other patients with this particular tumor type agreed to participate in such a trial. The most important goal of the first trial was to determine whether the new therapy would be toxic to patients. After all, the Her-2/neu protein is found on many normal cells, too, though in much smaller quantities than on tumor cells. Would interference with the normal copies of the protein cause problems? Slamon banked on the fact that malignant tumor cells carry many more receptor molecules than normal cells do. He hoped, therefore, that the tumor cells would respond much more strongly to interference with those receptors.

For safety reasons, very low doses of the antibody were given to the first patients in the study, doses far lower than the concentrations that affected tumors in mice. Those earliest trials showed two results: that the antibody did home in on tumors and that it seemed not to cause problems with ordinary cells that carry normal amounts of the Her-2/neu protein. As McNally enters the trial, Slamon is beginning to administer doses equivalent to those which stopped the growth of tumors in mice. And there is encouraging preliminary evidence that the experimental treatment may halt the progression of metastatic tumors in humans as well.

As the time for the treatment approaches, McNally insists that she feels hopeful, calm, and comfortable. "The nurse said she could tell I was scared because of my blood pressure," she demurs. "I don't feel it." Once the actual therapy begins, the process is physically so simple that it seems almost anticlimactic. McNally sits on a hospital bed for about ninety minutes, chatting with her husband and a nurse as the antibody drips slowly into one of her veins. Periodically she closes her eyes. "I don't want to just sit here while the antibody is going in," she explains. "I think it's more powerful to visualize it as it goes into my bloodstream and starts traveling toward the cancer cells."

The only sensation she describes is a slight itching, a common response, according to Slamon. McNally is certain that it bodes well. "I feel a real itching sensation in the area on my chest where the cancer is located," she reports, "and the itching is located only in that area, nowhere else. And it's pretty intense. When I'm doing the visualization, it's all working together because this is the only area of my body that I feel there is cancer in. So it is a good sign to me to have this whole area reacting." Later, she is given a small amount of radioactive antibody used as a tracer to allow Slamon to follow the protein on a monitor after it has entered her body.

If these trials go well, Slamon dreams of a more aggressive approach. Application of the antibodies themselves seems to interfere with the excessive growth signals that all the extra copies of the Her-2/neu protein are sending to the interior of the tumor cells. But Slamon wants to do more than slow down the renegades; he wants to kill them. One way to accomplish that is to attach much higher doses of radioactivity to the antibodies, which then become lethal "smart bombs," binding to cancer cells that display many Her-2/neu receptors and destroying them with a tightly targeted dose of radiation. Another option would be to attach a poison to the antibody, one researchers hope would be delivered in lethal doses only to tumor cells bearing all the extra copies of the receptors.

There is one final—and fascinating—complication to the story. Because the therapeutic antibodies were made by cells from a mouse, human immune systems respond to those antibodies as they would to any other foreign protein. After the first few treatments, therefore, McNally's body—and that of any other human patients—would mount attacks against them, disabling the potentially lifesaving molecules before they could accomplish their task.

To get around this problem, Genentech, using several techniques we will discuss in the next chapter, is producing what they call a humanized version of the antibody. The result will be a protein whose business end—the part that attaches to the Her-2/neu receptor—is made by a stretch of mouse DNA and whose body—the rest of the molecule—is produced by a piece of a human antibody gene. The result of this elegant gene splicing, which would have been impossible even a few years ago, just may be the first in a series of new, and hopefully much more effective, treatments for breast cancer.

• • •

IT IS TOO EARLY to tell whether this highly experimental treatment will eradicate Connie McNally's tumors, but preliminary results have been almost as positive as she could have wished. After the first trials turned up no toxic effects from therapeutic doses of the mouse antibody, subsequent trials employing the humanized version of that protein combined with more standard chemotherapy began. To everyone's delight, McNally has so far responded well to that combination therapy.

McNally and others who share her particular type of breast cancer are benefiting from Slamon's skill—and luck—in finding a previously identified oncogene in their tumor cells. With its product in hand, roads to many potential therapies are open.

But what of other types of tumors? The complexity of the molecular mechanisms controlling cell growth and differentiation ensure that many different kinds of mutations can cause runaway cell division. So it isn't surprising that oncogenes such as Her-2/neu, which function like damaged accelerator pedals, are only one cause of cancer. Researchers looking for other mutations, which have not been linked to known cancer-causing genes in animals, face a daunting task: to locate a single stretch of DNA whose base sequence is unknown among the 3 billion base pairs, 100,000 genes, and 23 chromosomes of the human genome.

Once again, Robert Weinberg's laboratory led the chase, this time in collaboration with Stephen Friend of the Massachusetts General Hospital Cancer Center and Thaddeus Dryja of the Massachusetts Eye and Ear Infirmary. They were the first among a half dozen competing teams to identify and isolate the first cancer-causing gene of a different type, a gene that causes a rare but deadly form of eye cancer known as retinoblastoma. This cancer forms on the retina of the eye, where specialized cells of the nervous system transform light into neural impulses. Retinoblastoma, known as a fatal condition for centuries, has always been rare, striking roughly one in 20,000 children. Extremely virulent retinoblastoma tumors were invariably fatal until advances in optical diagnosis and surgery made it possible to detect them early enough to save the patient by sacrificing the affected eye.

Once people stricken by retinoblastoma began surviving to adulthood and having children of their own, an unexpected and fascinating phenomenon emerged: children of many retinoblastoma survivors had a 50 percent chance of acquiring this otherwise extremely rare cancer. However, the disease also appeared in children with no family history of the condition. The mystery deepened. Finally, as the molecular revolution gathered momentum in the 1970s, several researchers seized upon retinoblastoma as an important target for genetic research. To summarize a long and challenging research project in deceptively simple fashion, researchers first guessed and then proved that the retinoblastoma gene is a damaged tumor suppressor rather than an oncogene, a malfunctioning "brake" on cell division rather than an accelerator. Each cell is normally equipped with two sets of identical molecular brakes controlled by this gene. A single functioning copy is enough to prevent runaway cell division, but if two copies are damaged, cancer results.

A child who begins life with two working copies of the retinoblastoma gene has reasonably fail-safe protection. Such a child must acquire, strictly by bad genetic luck, a mutation in both copies of the retinoblastoma gene *in the same cell* in order to be stricken with the disease. The chances of this happening are slim; only one child in about 30,000 develops the disease this way. Children who inherit a defective copy of the gene, on the other hand, live with the sword of Damocles hanging over their heads, because they begin life with only one functioning brake—only one protective hedge against this particular form of cancer. These children have a 95 percent chance of developing the disease. Why? Every cell in their bodies starts out with one defective copy of the gene, so if they lose the other copy in *any* cell, a tumor can develop. The chance that an error in DNA replication will damage the single remaining copy in at least one cell is quite high. In fact, many patients who carry a faulty retinoblastoma gene develop multiple tumors.

Fortunately for both researchers and modern-day patients, retinoblastoma has a feature that enabled researchers to locate it far more easily than most other genes. In these tumors, the genetic damage that inactivated the gene was actually visible: a portion of chromosome 13 was clearly missing. This was far from proof of the gene's identity, but it was a promising lead. Ultimately, Weinberg's collaborators isolated the gene and identified it. Soon thereafter, another research team deciphered its sequence of bases.

As a result of these discoveries, physicians can determine if an infant, or even a fetus, has or has not inherited a defective copy of the gene. Unfortunately, knowledge about the location and function of this gene's protein product (it resides within the nucleus and regulates a particular step in cell division) has not led to either preventive or curative therapies. But by identifying children at risk for the disease, physicians and parents can be put on full alert to have the child's eyes inspected carefully and regularly. With close surveillance, doctors have a very good chance of catching tumors when they are still small enough to be destroyed by laser surgery. That kind of intervention can often save not only the life of the patient, but his or her sight as well.

• • •

THE RETINOBLASTOMA GENE, although it was by no means easy to find, at least occurred in individual patients whose damaged chromosomes provided a visible clue to the malfunctioning stretch of DNA. Unfortunately, many genes that predispose their bearers toward certain cancers—like the type of breast cancer that plagues the Santi family— are not so obliging; their mutant forms give no visible evidence of their presence. Finding that type of gene is much more difficult in humans than it is in yeast or mice. Hunting for genes in animal models usually involves such techniques as exposing experimental subjects to carcinogens and breeding those individuals intensively for generations. Obviously, neither of those approaches can be considered with human subjects.

Instead, to find a human gene researchers have two prerequisites: a genetic condition that is passed from generation to generation and large families which display that condition. Ideally, such families should have many children, and surviving family members should span at least three generations. Those requirements send all human gene hunters to the far corners of the earth in pursuit of suitable families. One such search sent geneticist Nancy Wexler of the Hereditary Disease Foundation to Lake Maracaibo, Venezuela, where she had the good luck to find a huge family with the bad luck to carry the gene for Huntington's disease. Pedro León of the University of Costa Rica pursued his work in Cartagos, a town in which living descendants of an eighteenth-century immigrant led the way to a gene for inherited deafness. And Ray White of the University of Utah has assembled invaluable genetic data on Mormon clans, whose carefully kept genealogies and penchant for large families make them a geneticist's dream come true.

That's how the Santi family eventually crossed paths with Mary Claire King, a geneticist at the University of California at Berkeley who has been searching for a hereditary breast cancer gene for fifteen years. King brings to her work a powerful combination of intellectual curiosity and personal passion. Her best friend from childhood died of cancer at the age of thirteen, and a teacher and close friend recently succumbed to the disease. These personal experiences reinforced epidemiological evidence that advances in cancer treatment during the thirty years between the cases hadn't made much of a difference. "You have to do what is important to you," King reflects, "and you have to do what is fun for you. It's fun for me to solve puzzles, and this is a terribly important puzzle. As a woman, as a mother of a daughter, as a friend of other women, it matters to me enormously that we be able to do something to improve women's health."

King's persistence and dedication proved invaluable when, as a postdoctoral fellow, she began hunting a breast cancer gene. Back then the tools available for genetic analysis were so crude that her job was not unlike trying to use a pocket telescope to find the American flag planted on the moon by the astronauts. "I got involved with this work very naively," she recalls, shaking her head, "thinking that back in the mid-seventies it would be possible to address the genetics of breast cancer knowing what we knew then. Fortunately, we're not limited now to what we knew then or it would have been an incredibly stupid mistake!"

Holding up a picture of the twenty-three pairs of human chromosomes, King scans the black-and-white-banded images like an explorer gazing at the map of an unexplored planet. "We looked here, we looked here, we looked here," she recites, pointing each time to a different location. "We spent three years looking at this patch of chromosome 1," she reminisces, her voice not betraying the frustration she faced on recognizing that she had wasted precious time. As she continues her catalogue of blind alleys and false leads, the difficulty of her task becomes apparent; she and her team spent years searching in 182 different chromosomal locations that turned out to be wrong. Luckily, she didn't give up; the 183rd location was the correct one. She hasn't pinpointed the gene yet, but as this book went to press, she had narrowed it down to a small region on chromosome 17. Just how has King gone about hunting this gene? As she herself emphasizes, the work is now feasible thanks to the cooperation of families like the Santis and to a suite of recently developed molecular tools.

Finding the right families wasn't easy; none of those turned up by earlier investigators seemed to carry a gene for inherited breast cancer, so they weren't helpful to King. That wasn't too surprising, because only a fraction of all breast cancers are caused by any known inherited genetic defects. King and her team therefore resigned themselves to sifting through innumerable breast cancer cases before finding what they needed—large families with well-developed pedigrees that showed an unusually high percentage of premenopausal breast cancer.

Luck changed that situation dramatically. On the 125th anniversary of the National Institutes of Health, which funds King's work, a local television station taped a brief interview about her research. The next morning, Nancy Reagan's breast cancer was announced, and 126 network affiliates across the United States and Canada decided to air the interview with King. "Immediately we began to get thousands of phone calls," King recalls with glee. By the time the dust settled, they had so many promising leads that it took nearly eighteen months to follow up on them. Ultimately, the twenty families that participated in her work supplied the genetic material she needed for her search. One of those was the Santi clan, whose genetic counselor led them to King's lab.

ALFREDO SANTI's female descendants weren't surprised when King told them a few years ago that their family's breast cancer resulted from a defective gene. "I have always known something was wrong and that the cancer wasn't just random," admits thirty-one-year-old Dee Dee, whose mother, Iris, contracted the disease at thirty-nine. "It's no surprise I got my red hair from my mom. She may also have given me something else—this breast cancer gene."

"You take the bad along with the good," says Dee Dee's sister Katherine, who has already had a mastectomy because of her breast cancer. "Yes, we have this gene in our family, but we have this really great big loving, wonderfully supportive family. And that is probably the biggest reason you can get through all of it, because you have the love and support of your family."

These women now know that each one of them affected by cancer was probably born with one normal and one defective copy of the gene. Although every cell in their bodies carries that altered piece of DNA, it seems to cause cancer only in the cells of the breast and ovary. But because the gene is also present in egg and sperm cells, every child born to a parent with the gene has a fifty-fifty chance of receiving it. Many researchers think that this gene is a tumor suppressor gene, similar in action perhaps to the gene for retinoblastoma. But because this gene has not yet been fully characterized, its function and the way it causes cancer are still unknown.

Dee Dee has not developed breast cancer. She still doesn't know which copy of the gene she inherited from her mother. Has she inherited the normal gene and thus escaped the disease? Or does the defective one lurk, unseen, but ready to strike in the near future? But again, because this particular gene has not been identified, there is no way to tell for sure precisely what her status is.

"If I knew I had the gene, I would step up having children right away and breast-feed them," Dee Dee states resolutely. "Then I would go ahead and have my breast tissue removed. I know a lot of people think that is a rather severe course, but they haven't had to watch their mom and aunt go through what I have. And they haven't had to watch their sister get breast cancer at thirty-one." There are other reasons why the entire family has been eager to help Mary Claire King in her research. "I want some answers for my daughter," Katherine adds. "She's four now, but I want some answers before she's eighteen or twenty."

Participating in the study hasn't been difficult; all they've had to do is contribute blood for analysis. The white blood cells that are easily isolated from such samples, like the other cells in their bodies, contain all twenty-three pairs of chromosomes that comprise their genetic endowment. In King's lab, those isolated cells are provided with nutrients and treated to enable them to grow in culture indefinitely. They provide an everlasting source of DNA, just as a well-tended fruit tree can provide an abundant supply of peaches summer after summer. Any time DNA samples are needed, a batch of those cultured cells can be removed and broken open, releasing the collections of genes that are the basis of King's studies.

King's technique for finding the inherited breast cancer gene relies, in its early stages, on guilt by association. In police work, any person spotted at the scene of ten different crimes becomes a suspect, even if no one actually saw that person do anything suspicious. In the case of gene mapping, each "crime" is an occurrence of inherited disease, and the "suspects" are small bits of DNA randomly picked from the twenty-three human chromosomes. If a certain piece of DNA is *always* present in the Santi women who develop breast cancer, and *never* present in those who do not, that piece of DNA is suspect. Even if nothing is known about the function of the suspect segment, it and its immediate neighbors on the chromosome are searched in the hope that the killer gene is somewhere nearby.

But here any resemblance between genetic sleuthing and the glamorous work of movie private eyes ends abruptly. Gene hunting is a laborious and repetitive process. A scientist chops chromosomes into little pieces, chooses segments at random, and tracks the inheritance pattern of each segment through a family harboring their genetic disease of interest. That's what King and her team did with the DNA collected from their breast cancer families—they searched for a DNA segment that is inherited with breast cancer *all* the time.

If each scientist had to search all the hundreds of thousands of DNA fragments, at random, finding genes would be virtually impossible, except by blind luck. Imagine, for example, that you were asked to find a car broken down on Interstate 80 somewhere between San Francisco and New York. If that's all the information available, you have no option other than to cruise along the road until you get lucky. That's not an unreasonable analogy for the task geneticists face in hunting a disease gene; they must keep trying random chromosome fragments—cruising along the chromosome, in a sense—until they get the right one.

That's why geneticists need chromosome maps, and why providing markers that divide long stretches of DNA into small and identifiable segments is a primary goal of the Human Genome Project; as more and more DNA markers are placed on our maps of human chromosomes, gene mapping becomes faster and easier. One of the biggest complicating factors in gene hunts is the fact that, as noted in Chapter 2, chromosomes aren't passed from generation to generation intact. Instead, as parents' ovaries and testes make eggs and sperm, members of each chromosome pair swap pieces with their partners, so geneticists have to track genes as they are swapped back and forth on different pieces of the chromosomes between generations.

The solution to that problem is straightforward, though by no means simple. It works in some ways like the childhood games in which a search was guided by the hints "You're getting warmer" or "You're getting colder." If a geneticist finds that one particular marker—call it X—is inherited with a disease about 80 percent of the time, she can surmise that the marker is fairly close to her gene. She can then search all the other markers in the area. If the next marker she checks—Y, let's say—is inherited *less* frequently with the disease, perhaps 70 percent, she is getting colder—moving the wrong way along the chromosome. If Y is inherited *more* frequently with the disease, then she is getting warmer. The geneticist can thus use these DNA markers to zero in on her gene.

When Mary Claire King started her work, there were fewer than one hundred known genetic markers. Today there are several thousand, scattered throughout the human genome. Already, this increase in information has made gene hunts faster and easier. "What took fifteen years to do for breast cancer took two months for Pedro León to do with his deafness gene in 1992," observes King wistfully.

Using markers, chromosome maps, and families like the Santis, King has pinpointed her gene to an area roughly 2 million base pairs long on chromosome 17. With luck, its discovery could come any day; without luck, the last steps of tracking down the gene could take a long time. Once the precise location is known, scientists can recover the gene from that spot and finally identify it.

We can expect that at sometime in the near future, Mary Claire King and her coworkers will gather excitedly around a computer screen, watching as the sequence of the gene unfolds before their eyes. Yet what good will it do? Can this information help any breast cancer victims live longer and better lives? Researchers hope that once the gene is found, its protein will turn out to resemble something that has been identified in fruit flies or rats. If that happens—and if the role of that particular protein in regulating cell division is understood—the discovery might pave the way toward the development of a specific treatment aimed at this condition. That would place another class of breast cancers in the hopeful, though by no means guaranteed, situation of Slamon's treatment of cancers involving the Her-2/neu gene.

It is also possible, however, that the gene produces a protein whose role is more like that of the retinoblastoma gene, whose discovery has not yet paved the way toward new treatments. Women known to carry this gene who decide not to follow Dee Dee's radical course would at least know to monitor themselves carefully and continually, improving their chances of detecting the tumor when it is still at a very early, and therefore more easily treatable, stage. King has a hunch that this gene, like the retinoblastoma gene, may play a role in noninherited cancers as well. If that is the case, its discovery could offer the benefits of early diagnosis to a much larger number of women.

There is enormous potential benefit in having a genetic test for breast cancer, because it is far more difficult to detect small tumors in the breast than in the eye. "To find a malignant breast tumor now," King explains, "we use either mammography or a physical exam or both. And they work well. But to pick up a tumor by mammography, it has to be at least as big as the tip of my little finger. By the time the tumor is that big it has already undergone thirty cell divisions, and there is a good chance that some cells will already have broken off and invaded other parts of the body. By the time that little tumor undergoes just ten more cell divisions, it will weigh two pounds and will, of course, long since have killed the woman who carried it."

But if doctors could remove a few breast cells and detect the presence of corrupted genes, even in a single cell, they could detect breast cancer in its infancy. One researcher likens the rationale behind this approach to that of the much-debated Star Wars defense project, in which satellites detect the first flash of a missile being shot from the earth and destroy the rocket soon after it leaves the ground. If we could see the "first flash" of breast cancer—the first genetic alterations that cause the disease—we could remove the offending cells when they were mere molehills, not mountains. This technology, though not yet in hand, is feasible through logical extension of current procedures. Such early diagnosis would enable us to save many women's lives, and perhaps their breasts as well.

• • •

WEINBERG, SLAMON, KING, and their colleagues concentrate on oncogenes and tumor suppressor genes, pieces of genetic information that normally control cell division and which, if damaged, can cause cells to grow out of control and form a localized mass of cancerous cells. For such a cancer to become life threatening, however, some of those cells must "learn" to perform several specialized and intricate functions. First, they must work their way through the tough membrane that separates most young tumors from the circulatory system that could carry them around the body. Then they must survive in the bloodstream, work their way back out of the bloodstream, invade new tissues, and stimulate the growth of new blood vessels to feed them.

That's one reason why metastasis is still largely a "black box" from a genetic standpoint. In the opinion of Pat Steeg of the National Cancer Institute, it is likely that these diverse activities are controlled by several different genes. Yet Steeg, working with a devotion rivaling that of Mary Claire King, is determined to shed light on metastasis, because it is the main worry of anyone diagnosed with a tumor. "When a woman is diagnosed with breast cancer," Steeg observes, "the first three words she utters are 'Has it spread?' That is metastasis."

Steeg describes her work as a labor of love, adding that she savors "coming into the lab in the morning and seeing what is cooking." Yet this lighthearted enthusiasm is mixed with a strong dedication, tempered in the heat of personal experience. "I have had family members who have developed cancer," she recalls sadly. "I once had cancer myself. I have heard the word *carcinoma.* I can tell you that it is a horrifying experience. And I take breast cancer personally. I am a woman—one in nine women get this disease. It gives you a dedication that goes over and above the normal job requirements."

After years of hard work, her dedication may have paid off, for Steeg has discovered a protein that suppresses metastasis in animals. The way this protein showed up points out the value of both careful observation and a widely inquiring mind.

Steeg had several different tumor cell cultures in her laboratory, all of which could form local tumors when injected into mice. Some went on to metastasize, forming secondary tumors in different parts of the body, but some did not. What made the metastatic cells different from the nonmetastatic ones?

To find out, Steeg put pieces of DNA from the nonmetastatic cells into the metastatic ones to try to pinpoint the piece of DNA that could prevent metastasis. Eventually she located a single gene which could do just that. She called it *NM-23,* because it was the twenty-third gene from *n*on*m*etastatic cells that she examined. Subsequent studies showed that all normal body cells have working copies of this gene, while at least some metastatic tumor cells have somehow lost it. Although many other prominent researchers are skeptical, Steeg believes that this gene somehow prevents normal cells from forming metastases.

From a slightly different point of view, the NM-23 protein acts rather like a policeman for the body, apparently obliging cells to act like law-abiding members of the community. Responsible group behavior, of course, is a necessary characteristic of all the cells in virtually every multicellular living creature, from sponges to chimpanzees. If we assume for the sake of argument that Steeg's hypothesis about NM-23's function is correct, it isn't surprising to find that this gene—like a number of oncogenes and tumor suppressor genes and many other stretches of DNA —has been identified in a wide range of animals. Steeg is particularly impressed by the close resemblance of the NM-23 protein to a molecule in slime molds, peculiar organisms that last shared a common ancestor with humans more than a billion years ago.

Slime molds are intriguing by virtue of the fact that their cells behave in radically different ways at different stages of their life cycle. In one stage, each cell exists as an individual, crawling independently over rotting vegetation or damp soil with amoebalike motions. When food becomes scarce, however, the cells crawl toward one another and clump together, eventually forming a mass that looks rather like a slug. The slug then crawls around as one unit, leaving behind a trail of slime that gives this animal its name. Eventually, the slug enters a reproductive phase that produces another generation of independent cells.

What interest could a cancer researcher have in such creatures? Steeg learned that when slime mold cells are behaving as free-roaming individuals, they contain the NM-23 protein in very small quantities. But when the individual cells aggregate and begin acting as a single organism, the NM-23 protein is present at very high levels. Thus, she suspects that this gene product has some role in coordinating the activities of once independent cells. If she is correct, its absence allows—or obliges—cells to go their own ways; its presence obliges—or enables—them to work together as a unit.

"We then wonder," Steeg explains, "if the same process is going on in cancer. That is, when the NM-23 gene is working, the cancer cells tend to stay together as a formed tissue. But when NM-23 is lost, each of these cells goes its own way, and this is what we call metastasis."

Steeg used a molecular biology trick to effectively silence the NM-23 gene by preventing its messenger RNA from connecting with a ribosome to make protein. To do that, she began with the gene's DNA sequence, using its information to deduce the base sequence of its messenger RNA. With this knowledge, she assembled what is known as an *antisense* construct: a piece of DNA which makes messenger RNA that is exactly complementary to normal messenger RNA. When the two complementary RNA strands meet, they bind to each other, so neither can reach a ribosome and direct protein synthesis. Unable to make its protein, the gene is effectively silenced.

The results of this experiment were striking. Normal cells grow in petri dishes in the lab, but they stop dividing after they form a thin layer that covers the surface of the plate. But cells in which the NM-23 gene was silenced in Steeg's experiment virtually grew right out of the dish. More important, Steeg found that this gene seems to be significant in at least certain forms of human breast cancer. Tumors that contain large amounts of NM-23 protein, it turns out, are significantly less likely to metastasize than tumors with very little NM-23 protein. "What this suggests," Steeg observes excitedly, "is that NM-23 regulates the same kinds of interactions in the cells in our body. When these cells are together in an organized tissue, NM-23 is expressed, but when they lose this organization and it is every cell for itself, we lose NM-23. It's an amazing correlate."

Steeg hopes that this line of inquiry will lead to new therapies for women with metastatic cancers. She has already found that putting the gene into human breast cancer cells in mice drastically reduces their potential to form metastases. She is working with French Anderson of the National Institutes of Health, a human gene therapy pioneer, to develop NM-23 gene therapy for women with breast cancer.

• • •

THE STORIES INTRODUCED here represent but a few of many parallel and converging lines of research that employ similar modes of inquiry: the induction of cancer in experimental animals, mapping genes for inherited cancers, and constant cross-checking with basic researchers studying genes that control cell division. Molecular biologists using these techniques seek to understand cancer's ultimate origins in DNA. Their work is guided by the faith that in unpredictable and unexpected ways, the information they uncover will some day aid in diagnosing and treating this fatal disease.

For a decade or more, from the middle 1970s through most of the 1980s, their efforts remained largely an act of faith. In the past few years, however, their labors have begun to pay off. In addition to the promising leads reported here, work on the molecular biology of colon cancer has already pinpointed one gene strongly associated with that disease. In addition, various research teams have found abnormalities involving the retinoblastoma gene in all small cell carcinomas, a common form of lung cancer, most sarcomas, one-third of bladder carcinomas, and more than 20 percent of breast cancers. These findings emphasize two points: that genes found in a rare form of cancer may also appear in more common forms, and that understanding a finite number of such genes may ultimately prove to be of great clinical value. Many researchers and clinicians share what a few years ago would have been dismissed by many as an unrealistic hope: the belief that a new series of effective cancer treatments may be available within a few years of the millennium.

No matter how effective cancer treatments may become, however, it will always be safer and cheaper, not to mention less traumatic, to prevent cancer than to cure it. Unfortunately, certain aspects of contemporary lifestyles in developed countries seem to be working against us. We are eating richer food, delaying childbearing, and living longer—and we consider these changes to be improvements in our quality of life. Yet research has shown that all these modifications in habits increase our risk of ultimately contracting certain types of cancer.

Diets high in fat, particularly animal fat, for example, seem to cause a more rapid turnover in the cell layers lining the colon. It isn't clear precisely *why* a more rapid rate of cell turnover should increase the rate of genetic damage which causes colon cancer as much as it seems to, but studies suggest that reducing animal fat consumption by 50 percent could cut colon cancer risk in half. Further reductions in intestinal cancer can be expected from other dietary changes as simple as decreasing alcohol intake and increasing fiber consumption, although the molecular mechanisms for these phenomena are not at all clear.

Certain realities of cancer are even more clear cut. Epidemiologists estimate that one-third of all cancer cases in the United States could be avoided entirely if smoking were abolished; tobacco has been linked to cancer not only of the lung, but of the bladder, esophagus, pharynx, pancreas, and kidney. On the prevention side, regular checkups, particularly for breast and prostate tumors, make early intervention possible and greatly increase survival rates for those types of cancer.

No matter what preventive measures we take, however, a significant number of cancers will still occur in elderly individuals as the apparently inevitable consequence of aging. It is clear that as our molecular understanding of cancer grows, hopes for successful treatment of all forms of cancer will grow as well.

5

Cell Wars

. . . our townsfolk . . . disbelieved in pestilences.
A pestilence isn't a thing made to a man's measure;
therefore we tell ourselves that pestilence
is a mere bogey of the mind, a bad dream that will pass away.
But it doesn't always pass away,
and from one bad dream to another, it is men who pass away.

Albert Camus,
The Plague

AT THE CENTERS for Disease Control (CDC) in Atlanta, deep in the bowels of a building protected by armed guards, a man in hospital scrubs opens a steel door and enters a small anteroom. After the door seals behind him with a solid, reassuring *thunk*, he dons what looks like a Buck Rogers space suit, complete with helmet, gloves, and boots. He maneuvers through two more hermetically sealed doors, enters a laboratory equipped with centrifuges, autoclaves, incubators, and glassware, and sets to work —reading genes. Why the elaborate security for what seems to be a strictly run-of-the-mill molecular biology laboratory? Because this researcher is part of the team sequencing the genome of one of the most feared mass murderers in history, the *Variola* virus, commonly known as smallpox.

Americans today know smallpox only through books and films, because a global program mounted by the World Health Organization (WHO) seems to have eradicated the disease. As far as we know, variola virus survives only in culture at two maximum-containment laboratories: the American facility at the CDC and a similar institution in Moscow. Once the 175,000 base pairs in the variola genome are known, extant laboratory stocks will be destroyed, and the virus will become the first species humans have intentionally driven to extinction for good reason.

When the last variola particles are gone, all that will remain of the threat they once posed will be a sequence of bases recorded on computer disk.

In the meantime, each year at the end of January, Charles Whitaker of the Parke-Davis flu vaccine production unit is busy—very busy. Unlike many production lines, his isn't hustling just to make money. Whitaker, his staff, and other vaccine producers are racing against time to save lives. They are rushing to prepare 30 million doses of influenza vaccine, each of which confers protection against three different forms of the virus, for distribution across the country the following fall. At first glance, their urgency seems exaggerated. Isn't the flu just a minor inconvenience? A week with a cough and a runny nose? But the flu's innocence is deceiving; it costs the United States a billion dollars annually for doctors, drugs, and hospital care, and an additional \$3–\$5 billion each year in missed work and lost productivity. As many as 20,000 people in the United States alone die of influenza during an average winter.

There's a larger threat as well. Periodically—every ten to thirty years on average—a particularly virulent strain of flu appears. In 1918, one such strain caused a global flu epidemic that killed at least 20 million people, though some experts argue that the actual death toll was far higher. "That was the worst flu epidemic in human history," declares Robert Webster of Saint Jude's Hospital in Memphis, Tennessee. "It really zapped the world. The Germans blamed the Allies and the Allies blamed one another, particularly the Spanish, so they called it the Spanish flu. It actually finished World War One." How did a virus end the war? It killed so many German soldiers that their last major offensive was severely crippled.

Less fatal, but equally widespread flu epidemics occurred in 1957, 1968, and 1977. Another could occur at any time, and there is no guarantee that the next one won't be as serious as the 1918 version. That's why each winter finds flu researchers and vaccine producers scurrying to identify new strains and to get vaccines ready on time.

At first glance, the contrast between these two scenes seems paradoxical. The virus that causes flu, which for most of us means little more than a few nights of headache and fever, keeps the medical establishment on red alert year after year. Yet the virus that causes smallpox, a scourge of human populations around the world throughout recorded history, seems to have been vanquished. Why is the flu virus more adept than variola at eluding medical science? Why does a vaccination against smallpox protect for up to a decade, while a flu shot offers only partial protection for only a single season? And why are some flu strains nearly harmless while others cause life-threatening disease?

Finding answers to these questions leads us to a molecular battleground on which war has been raging since life began. Humans, like all animals and plants, live under constant threat of attack by microscopic organisms called pathogens, which fight to make a living in our bodies at our expense. Campaigns in the ongoing struggle between pathogens and hosts are waged at the molecular level, the outcome of each skirmish depending on the relative strength of opponents' genetic intelligence. In a very real sense, the battle is a contest between their genes and ours.

Viruses, in fact, are little more than tiny bags of genes, a few concise genetic "sentences" written in either DNA or RNA and wrapped in a "coat" of protein. They are so simple that they aren't alive in the usual sense. Lacking the cellular machinery needed to perform *any* of life's functions on their own, viruses survive by acting like renegade genetic engineers and taking over the inner working of host cells. After a virus "tricks" a host cell into taking it inside, some of its genes turn host genes off, others compel the host's molecular machinery to copy the viral genome, and still others induce host ribosomes to manufacture viral coat proteins. Most viruses ultimately kill their host cells, at which point hordes of infectious particles break out in search of new victims. In the process, they cause diseases ranging from common colds and flu to measles, herpes, smallpox, and AIDS.

Other pathogens that afflict humans include bacteria, which cause tuberculosis, pneumonia, and plague; a variety of single-celled parasites, which cause traveler's diarrhea, malaria, and sleeping sickness; and many different kinds of "worms," which produce debilitating and often fatal conditions including schistosomiasis. These organisms don't usually engage in direct genetic takeovers of host cells; instead they use instructions coded in their genes either to consume our living tissues or to drain us of vital resources. What's more, most pathogens disrupt our physiology in ways that cause what we recognize as symptoms of disease: coughing, sneezing, open sores, or diarrhea. As they discomfit us, these symptoms are pathogens' tickets to new hosts: sneezing broadcasts influenza virus; diarrhea spreads amoebas that cause dysentery; bites from mad dogs spread rabies virus.

All these pathogens' genetically programmed survival strategies have been refined and improved over eons of evolutionary time. Their deadly efficiency has affected human history to an extent that is difficult for us to comprehend today, raised as most of us have been in a world where lethal epidemics are aberrations. Nevertheless, when the Black Death swept Europe between 1347 and 1351, roughly one-third of the population perished. Subsequent visitations of plague claimed nearly a third of the inhabitants of Norwich, England, in 1579, half the Milanese in 1630, and nearly two-thirds of the Genovese in 1656–1667. Several historians agree that smallpox and other viruses, rather than guns and horses, were the real reasons for the decimation of indigenous populations throughout the New World following contact with Europeans.

For most of history, science could offer no explanation for such plagues, so superstition filled the gap. Both Aztecs and Spaniards attributed smallpox epidemics to the superiority of the white man's god over local deities. English settlers in Virginia ascribed their heathen neighbors' death to divine retribution for acts of witchcraft against Christians. In Europe, epidemics were seen as God's punishment for sin, particularly sins of the poor, minority groups, foreigners, and other scapegoats on society's fringe. Gypsies, Jews, and pagans, among others, were ostracized, executed, or incited to self-abuse. Such practices were fueled by denial; blaming plagues on deviant "others" locates a moral cause for suffering among the "guilty" and distances the "innocent" behind illusions of safety.

Scientific approaches to disease weren't totally lacking, although Western scientists didn't really begin to associate newly discovered microorganisms and polluted water with disease until the second half of the nineteenth century. Centuries earlier, however, physicians in China, the Middle East, and Africa noticed that people who survived mild cases of certain diseases were somehow protected from more serious bouts with the same illness, often for the rest of their lives. Building on this experience, local physicians developed strategies for exposing patients to smallpox and certain other diseases in a limited way, in the hopes of inducing that same mysterious protection. Some of these practices seem to have been reasonably successful, at least if their hit-or-miss techniques didn't accidentally cause full-blown disease.

The concept of immunity (from the Latin *immunis*, for "exempt") was brought to European attention in the late 1700s by British physician Edward Jenner. He noticed that milkmaids who recovered from a mild disease called cowpox seemed immune to smallpox. Insight, intuition, and courage led Jenner to inoculate eight-year-old Jamie Phipps with fluid and pus taken from the cowpox lesions of a milkmaid. Jenner later tested

The milky substance adhering to this glass rod is DNA that has been released from cells broken open by detergent. Despite its nondescript appearance, DNA is the molecular "secret of life"; information it carries directs the vital functions of all living things, and traces its ancestry back to the dawn of life.

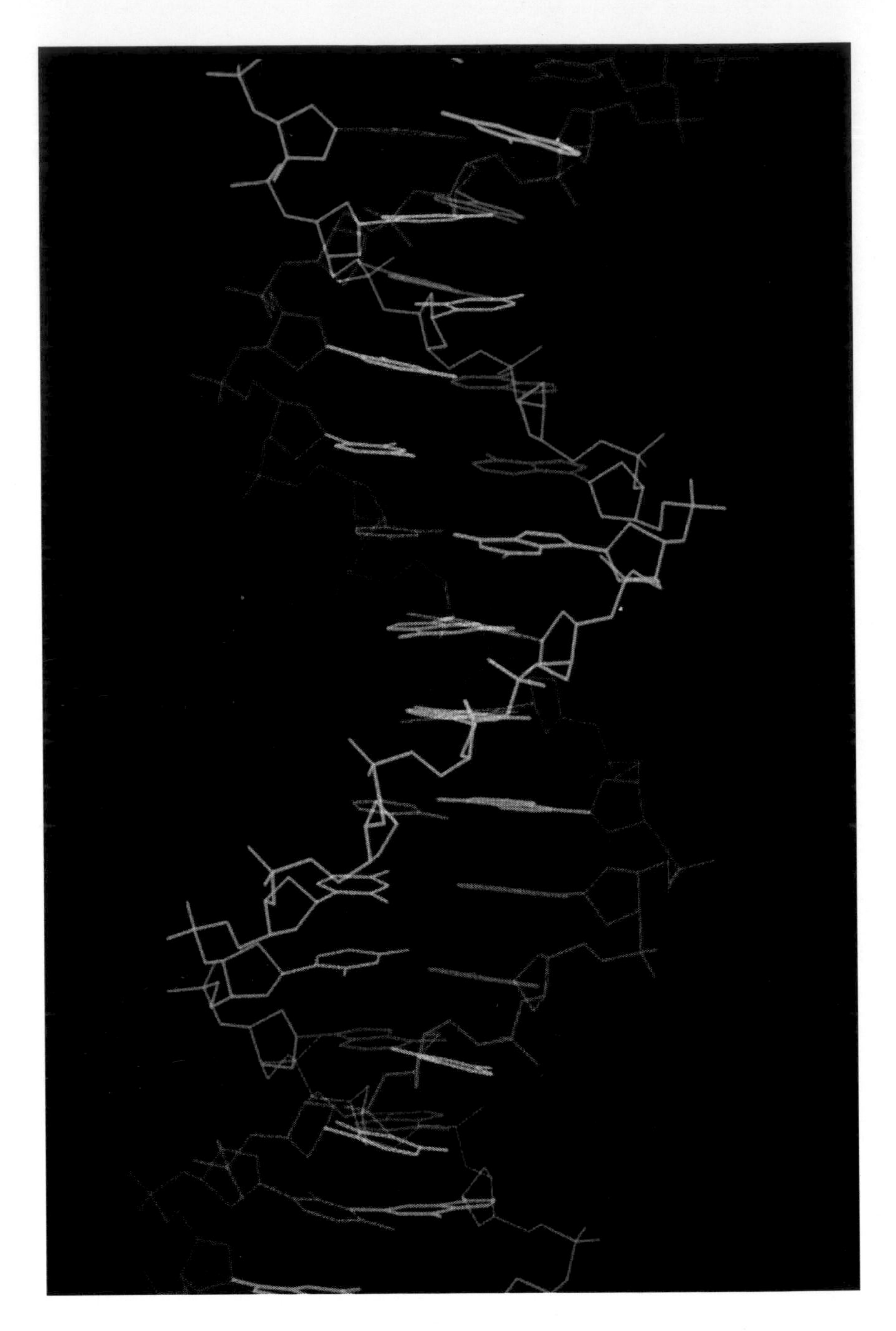

A computer-generated image of DNA reveals two chains whose coiled embrace creates a double helix. Each chain consists of a molecular backbone (one in red, the other in blue), along which are strung four bases: adenine (dark blue), thymine (yellow), guanine (green), and cytosine (purple). Bases pair up according to a strict formula—adenine with thymine and guanine with cytosine—forming steps in a spiral staircase that carries the code of life.

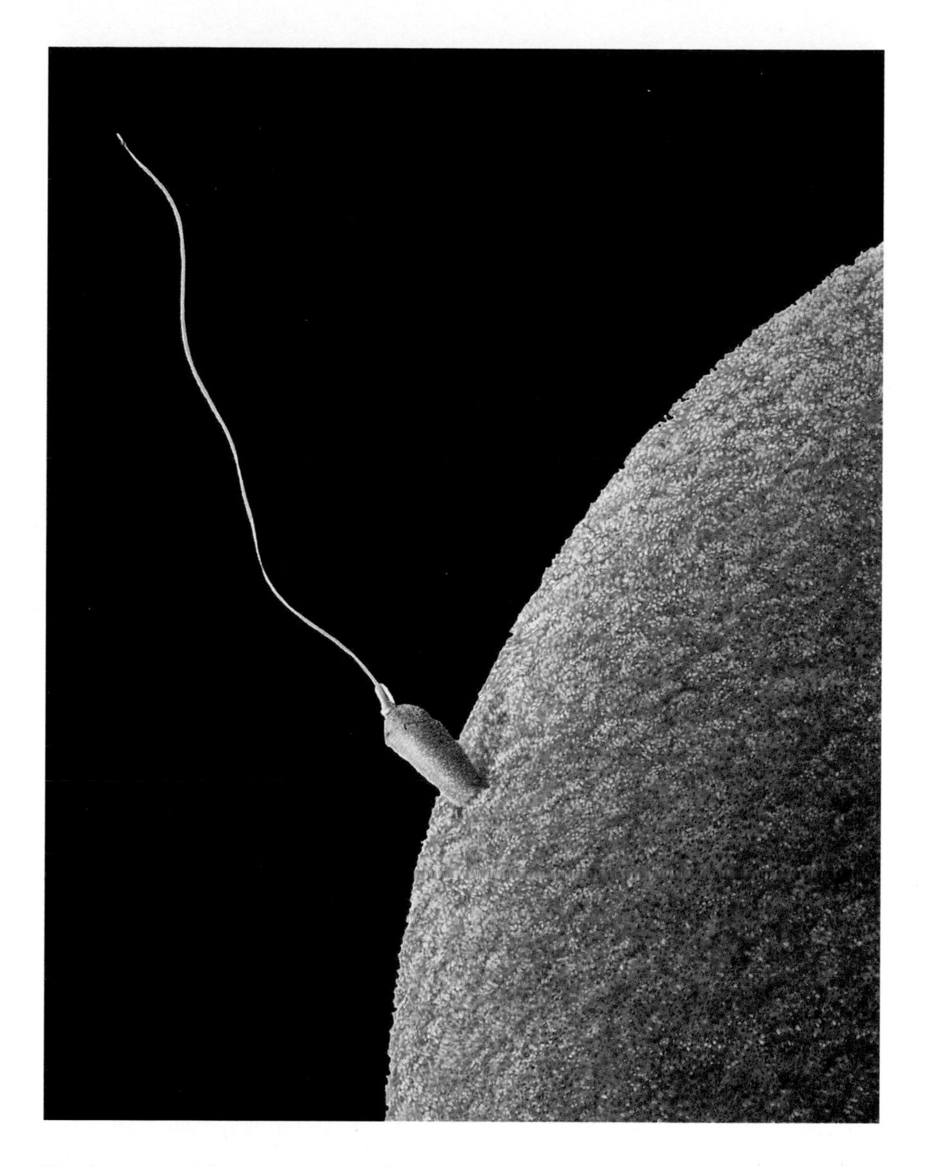

For better and for worse, every individual's genetic endowment is determined at the moment of conception. Sperm and egg each carry a random selection of parental genes. Their fusion creates a genetically unique individual.

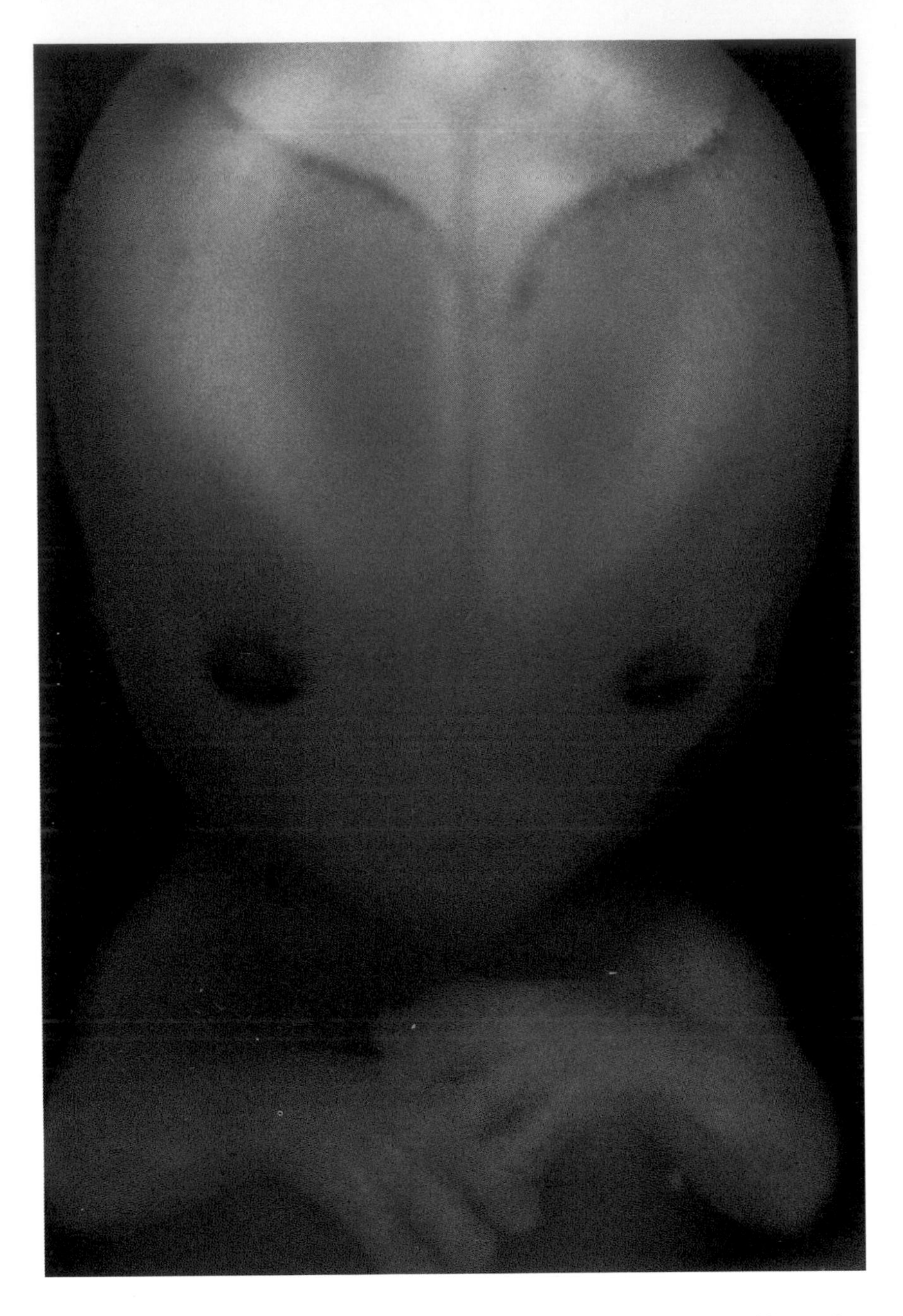

As an embryo develops, genetic master controls orchestrate the workings of countless genes, which determine sex and influence innumerable physical and physiological features. Some of the genes operating in this three-month-old fetus remain active in adult life. Others operate only during early development, and are normally silent thereafter.

Spanish athlete María José Martínez Patino, an XY woman, was disqualified from Olympic competition in the mid 1980s because, although she was biologically female, her unusual genetic constitution led to her being "legally" classified as male. Patino ultimately surmounted challenges and won the right to enter the 1992 Barcelona summer games.

Vital events in the molecular world are often tied to apparently innocuous situations in the macroscopic realm. In Yunnan, China, humans, pigs, ducks, and fowl live in close quarters. Each of these animals may be infected with its own strain of influenza virus. The spikes that cover these viruses, depicted here in a false-color electron micrograph (right), play key roles in infection and in triggering the immune response. Genetic intercourse among different virus strains—encouraged by close proximity of their hosts—may produce changes in the genes that control the spikes, unleashing deadly epidemics like the Hong Kong or the swine flu.

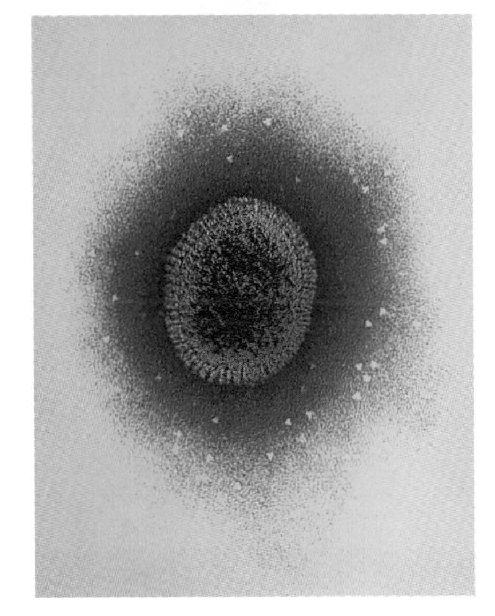

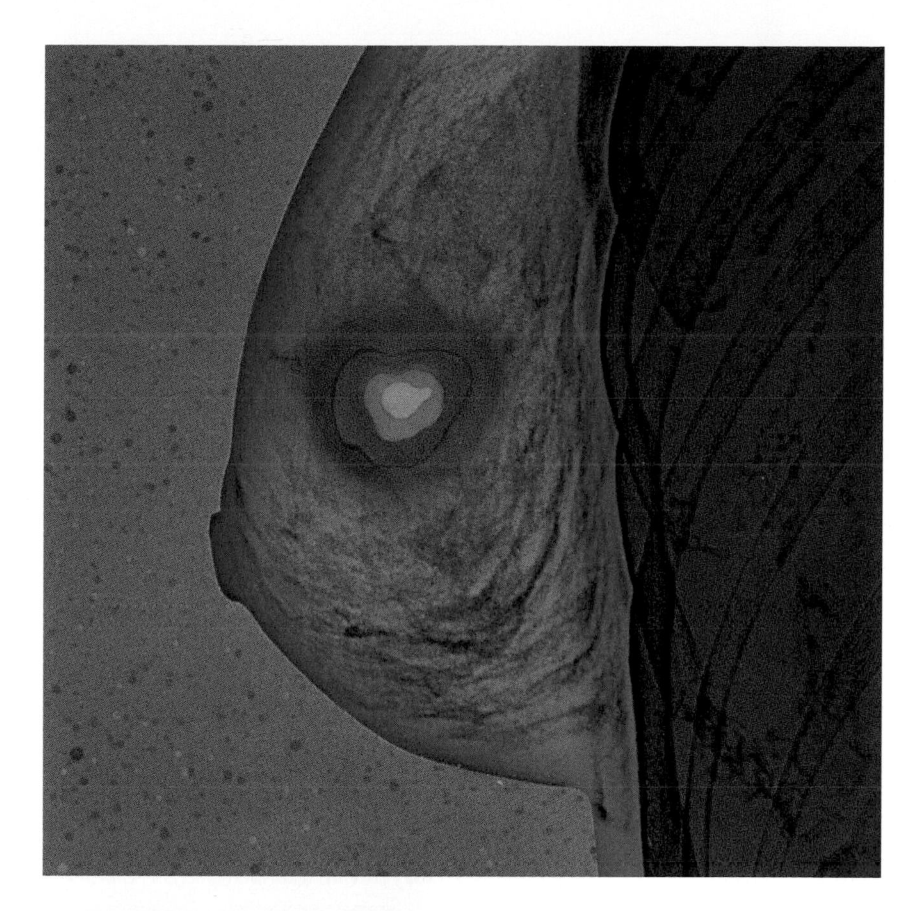

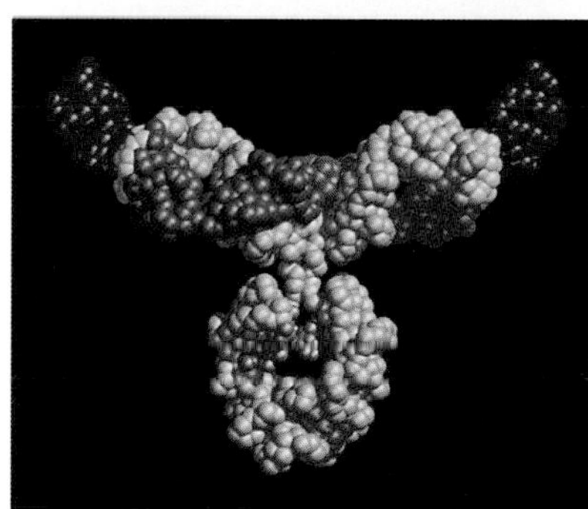

The glowing region in this false-color mammogram reveals a tumor, a potentially deadly collection of cancerous cells that, by acquiring mutations in vital control genes, evade normal constraints and threaten the life of their host. One new molecular weapon in the war against breast cancer is an antibody (left) tailored by genetic engineering to recognize and attack one particular protein—the product of the Her-2/neu gene—whose unusual abundance in certain cancer cells sets them apart from normal body cells.

Many modern Senegalese carry the same genetic protection against malaria that baffled early European explorers in West Africa, nearly all of whom died from "the fever." However, the gene responsible for this protection can be deadly: a double dose of it produces a variant blood protein known as Hemoglobin-S, which induces sickle-cell anemia. Red blood cells filled with Hemoglobin-S can distend into rigid sickle shapes (right) that lodge in the tiny blood vessels, leading to serious problems throughout the body.

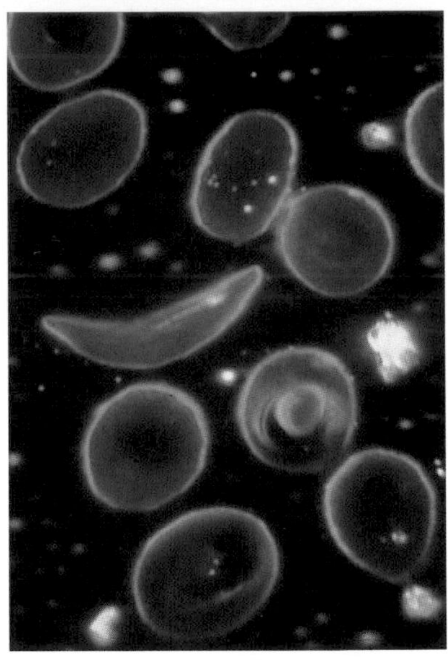

All these tropical tree snails are members of the same species. Yet the visible permutations in their colors and banding patterns are only part of the inheritable variation that makes each of them unique. That variability, like the changes in DNA that produce Hemoglobin-S and other physiological differences among us, is the molecular basis of evolutionary change.

The silverswords, more than two dozen native Hawaiian plants, evolved recently from a common ancestor. Thanks to small but powerful random changes in their genes, these plants display extraordinary diversity in form, size, and habitat. The most striking species is *Argyroxiphium sanwichenses,* shown growing under near-desert conditions in Maui's Haleakala crater.

The injection of foreign DNA into a mammalian egg engineers genetic changes far more quickly than natural selection, surmounting natural barriers that normally prevent genetic exchange between species. In this image from the cutting edge of genetic engineering, a mouse egg is held by suction against a micropipette. An even smaller pipette approaches, laden with artificially assembled DNA that carries a selected foreign gene.

A biological fermentation tank, part of a bioreactor complex, uses extraordinarily sensitive and sophisticated equipment to coddle genetically engineered bacteria and harvest their products.

This family of ewes represents the next generation of bioreactors—transgenic animals whose genetically altered mammary glands produce and secrete valuable proteins into their milk. Tracy (at center) is the transgenic mother of nonidentical twin daughters, who flank her. One daughter inherited a gene inserted into the fertilized egg that produced Tracy; the other daughter did not.

These youngsters were posed to dramatize their similarities and differences. Although we share roughly 99 percent of our genome with chimpanzees, the physical and behavioral traits that distinguish us trace back to the few genes that set us apart. How can so few genetic differences produce such strikingly different brains?

The identical twins were produced by the splitting of a single fertilized egg. They therefore began life with virtually identical genetic endowments and similar appearance. Yet even during infancy they developed characteristics that visibly distinguish them. To what extent are their *invisible* traits, including IQ, temperament, and tendency to develop behavioral disorders, rigidly determined by their genes, and to what extent are they subject to modification?

EUGENICS

A Journal of Race Betterment

Volume II APRIL 1929 Number 4

Hereditary Mental Aptitudes In Man

By HENRY H. GODDARD

DOES AMERICA NEED MORE MORONS?

Responses by

W. S. ANDERSON WALTER B. PITKIN JOHN C. PHILLIPS ELLSWORTH HUNTINGTON

30 cents the copy, $3.00 a year

PSYCHOLOGY NUMBER

The subtitle of this eugenics journal reminds us that ideas now considered repugnant were once scientifically—and socially—respectable. Advocates of human behavioral genetics still publish articles with titles not unlike the lead article featured on the cover of this issue; critics argue that modern studies suffer from many of the same flaws and biases that characterized those published more than sixty years ago.

his treatment by intentionally exposing Jamie to extracts from a smallpox lesion. Luckily—for both Jenner and Phipps—the procedure was successful, and vaccination became a central tool of Western medicine. The procedure, though, was literally a shot in the dark; neither Jenner nor his colleagues had any idea what caused cowpox or smallpox, and no one had a clue as to why the vaccination—as it came to be called, after the Latin *vacca* for cow—actually worked.

Today we know that Jenner was tinkering with the immune system, an elaborate, genetically guided network of cells that recognizes and attacks foreign invaders at the molecular level. On the most basic level, the immune system's job is simple. "When you say 'Who am I?'" notes Columbia University parasitologist Dickson Despommier, "biologically, the immune system tells you. You are as different from everyone and everything else as a fingerprint, and your immune system makes sure you stay that way. It keeps your biological integrity intact."

In more precise terms, the immune system learns to distinguish self (cells and proteins native to the body) from nonself (organisms and compounds from outside the body). It then attacks and destroys any "nonself" that it encounters. Each time pathogens enter our body, the offensive strategy programmed in their genes is pitted against the molecular defenses carried in our genome. Neither side's genetic instructions are immutable; over time, their genes evolve in ways that help them elude our defenses, while our genes respond with the genetic equivalent of escalating counterintelligence. Critically, once the immune system "remembers" any enemy it encounters, it remains alert to fend off recurrent attacks before damage is done.

What does this extraordinary defensive network look like? Because pathogens can attack the body anywhere, the immune system must be able to respond wherever and whenever threats to the body's integrity appear. It makes sense, therefore, that its most important components are several types of blood cells that circulate throughout the body. *Macrophages,* whose name means "big eater," cruise around the body searching for foreigners, which they engulf while sounding an alarm to other cellular members of the defense consortium. *B-cells,* the functional equivalent of munitions factories, respond to invasion by manufacturing and releasing protein "weapons" called *antibodies. T-cells,* often dubbed the foot soldiers of the immune system, engage in direct combat with the enemy. These cellular warriors come in two basic flavors, "helpers" and "killers," whose functions are indicated by their names.[2]

This molecularly guided cellular network keeps pathogens at bay so successfully that most of us—in industrialized countries, at least—spend most of our lives healthy, rather than lurching from one illness to another. That's part of what makes epidemics so terrifying; the outbreaks we designate as plagues occur only on rare occasions when the immune system fails on a grand scale. If there was ever any doubt that we owe our health largely to the vigilance of our immune system, it was laid to rest by the emergence of AIDS; when that disease destroys our defenses, we fall prey to parasites that normally don't even make us sneeze. The best that medical science can do to date is delay the inevitable.

That's why molecular biology is such an enormous boon to medicine; we haven't a clue as to how we could ever replace the immune system, so our best strategy is to understand it and learn to boost its efficiency whenever possible. Thanks to an arsenal of new molecular tools, researchers can study, and intervene in, battles between host and parasite directly at the level of interacting genes. Researchers are also learning ever more effective strategies for "teaching" the immune system to ward off new viruses and to defeat parasites that have evaded it in the past. To see how the cells of the immune system fight infection, and how molecular biologists help them in their work, we return to the Centers for Disease Control, where the work of preparing the following season's flu vaccine is in full swing.

• • •

"WHAT WE DO HERE at the CDC," Nancy Cox, chief of the Influenza Branch, tells us, "is receive viruses from virtually all over the world. We grow them to produce enough to test and see which ones are different." By way of explanation, Cox shows us a dozen rather ordinary-looking deep-freeze units. Their contents look inconspicuous enough—thousands of vials, each containing a few droplets of frozen, cloudy liquid and labeled somewhat enigmatically. One label reads "A/Beijing/98/89"; "B/USSR/100/83" is written on another. Each of these vials, we learn, contains a different strain of influenza virus, some of which were collected as long ago as 1933. The labels record where the samples came from and when they were collected.

Some of the old virus strains on file were nearly harmless; others caused serious epidemics in which thousands of people died. The new strains, which arrive by the hundreds each fall, could be either feeble or lethal. The job of Nancy Cox and her team is to figure out which is which—

in time to produce a protective vaccine before the new forms of the virus spread across North America.

Simply looking at the vials' contents, even under the most powerful light microscope, reveals nothing, because flu viruses are so tiny that two and a half million of them lined up next to one another would barely cover an inch. They can be so small because they carry minimal baggage—just under a dozen genes arranged on eight pieces of RNA and bundled in a protein coat. Under the electron microscope, the coat looks like a medieval mace, a ball studded with spikes that appear as though they could bludgeon their way through the toughest coat of armor.

In reality, however, the spikes on the flu virus, rather than bashing their way into our cells, serve as counterfeit "keys" that open "doors" to the cell; certain proteins in the virus's coat convince receptors on the cell surface to allow the intruder to enter. Once that molecular subterfuge smuggles a flu virus into a cell, viral genes quickly take control; within eight hours, a single virus can produce a hundred thousand copies of itself, destroy its host cell, and release enough viral progeny to wreak havoc elsewhere.

Unlike viruses that cannot survive outside the body for even a short while, flu virus remains viable for as long as twenty minutes in droplets of mucus or saliva. That's why it can be spread so quickly and easily. If a stricken Korean businessman sneezes in the crowded Honolulu airport, a virus he picked up in China a week earlier can be unleashed on New York, London, Brussels, and Paris within forty-eight hours. But transmission is only part of the influenza story. To understand fully why the CDC works so hard to collect flu samples from around the world, we need to know more about how the immune system attacks the virus and how the virus fights back.

WHENEVER WE'RE INFECTED by a flu virus our body hasn't met, the pathogen enters our cells and begins its dirty work. That first round of activity usually peaks within forty-eight hours, so we feel miserable in short order. But as increasing numbers of viruses circulate in the body, they spur the immune system into action. Some of those viruses eventually come into contact with a succession of B-cells, some of which recognize them as foreign and sound an alarm call. But how does recognition take place?

As noted earlier, B-cells manufacture antibodies, molecular machines programmed to recognize and attach to foreign proteins. Essentially, one end of each antibody molecule functions like a molecular machine whose program directs it to "find a particular protein and stick to it like glue." The other end of the antibody acts like a signal flag, announcing, "There's an enemy here; do something!" Once invading parasites are tagged by antibodies, the signal flags attract other components of the immune system.

As it turns out, the genes contained in each B-cell direct it to produce one particular antibody, which recognizes foreign proteins of only certain specific shapes. Each individual B-cell puts numerous copies of its antibody on its surface, where they act like the sort of cell-surface receptors discussed in Chapter 4.

The body contains an enormous number of different B-cells, each of which carries a different antibody that recognizes a different protein. In fact, there is likely to be at least one B-cell receptor in your body that can recognize almost any protein on earth. If that sounds incredible to you, you're not alone; when researchers first discovered antibodies and B-cell receptors, they tried repeatedly to find proteins that the immune system couldn't recognize. They failed. Even completely artificial proteins, compounds that had never existed, were always recognized. The ability to recognize virtually any protein is at once the most remarkable and the most basic characteristic of the immune response.

Recent studies estimate that the immune system of every human being can respond to as many as *10 billion* specific foreign proteins. Even immunologists consider that an astounding number. How is this possible? We've said that the entire human genome contains only about 100,000 genes, each of which directs the assembly of a protein. But if that's the case, how could the body make 10 billion different kinds of B-cell receptors alone? There aren't enough genes to go around. What's more, how could your genes, which your parents bequeathed to you at conception, "anticipate" the shape of proteins in a flu virus you won't pick up until nine months hence?

The answer lies in a strange and wonderful characteristic of certain B-cell genes. We said earlier that action of the "sticky" part of the antibody molecule could be described by a molecular sentence that says something like "Find a particular protein and stick to it like glue." Now imagine that in order for an antibody to grab on to a protein, its instructions have to be more specific. Instead of being told simply to "find a particular protein," or even to "find a particular *foreign* protein," it needs to be told almost exactly what the target protein looks like.

So we're back to the same problem: How can your genes know in advance about the shapes of foreign proteins you'll meet during your lifetime? They can't. Instead, as B-cells are produced by stem cells in bone marrow, the genes that direct the assembly of antibodies in each cell shuffle their parts around, combining and recombining them differently. As the cells containing these genes divide, each daughter cell receives a copy of the antibody-producing gene which is slightly different from every other copy. The result is millions of slightly different genetic sentences, each of which instructs the antibody it makes to chase down a slightly different kind of protein.[3]

The rearrangement works because the lengths of DNA that mature into antibody genes start out life carrying many more molecular "words" than are necessary to make a single protein. Most of those extra words are stacked in the part of the genetic sentence we've concentrated on so far, the part that says something like "Find a particular protein and stick to it like glue." There are so many extra words in that part of the gene that it doesn't really resemble a sentence. Instead, it seems more like an old-fashioned Chinese restaurant menu with three long lists from which you're meant to choose one from column A, one from column B, and one from column C.

We might imagine that every baby B-cell starts out with part of its antibody gene looking like this: "*Find a protein with a:* round, long, twisted, straight, curved, *piece that looks like a:* helix, loop, safety pin, tie clip, button, *at its:* beginning, middle, end, *and stick to it like glue.*" As the B-cell matures, it randomly selects one word from each set of alternate choices to make its own personal version of the gene. The parts printed in italics stay the same, but only one intervening word remains in each spot.

Thus, the final genetic sentence carried by one B-cell might read: "*Find a protein with a* **round** *piece that looks like* a **button** *at its* ***end*** *and stick to it like glue.*" Another B-cell that started out with the very same gene might end up with a final version that reads: "*Find a protein with a* ***straight*** *piece that looks like a* ***tie clip*** *at its* ***beginning*** *and stick to it like glue.*" A single, original stretch of DNA can thereby produce a wide variety of final molecular sentences.

The real antibody genes carry many more alternate words in each set than our simple example did. In one actual genetic set there are hundreds of alternates, in the second set there are somewhere around twenty, and in the third set there are four. You don't have to be a mathematician to appreciate that random combinations of that many different instructions can produce millions of different B-cells, each of which carries antibodies primed to respond to a different protein.

Most of those antibodies are useless at any moment in time, because most of them don't encounter proteins they recognize. But sooner or later, an invading virus is likely to bump into a B-cell whose antibody just happens to recognize part of its protein coat. Once that happens, the result is dramatic, for as soon as an antibody on the B-cell surface attaches to a foreign protein, it sends a molecular signal to the interior of the cell. As a result, the biochemical machinery of that particular B-cell, which has been floating around in the bloodstream not doing much of anything, roars to life.

Meanwhile, some T-cells, which carry the same sort of receptor molecules on their surfaces, are also activated by contact with foreign proteins carried by the virus. If an activated B-cell meets a similarly activated helper T-cell, the B-cell is stimulated to grow and divide rapidly, producing many, many daughter cells, each of which can produce antibodies targeted at the protein that stimulated the original cell.[4]

Many of the daughter B-cells quickly mature into dedicated antibody factories, each of which churns out numerous copies of the particular antibody it is dedicated to produce. Once these antibodies are released in large numbers, the viruses that provoked them are in trouble. The sticky ends of the antibodies grab on to the virus's spikes, while the other end flags them for destruction. Viruses that are "stuck" to antibodies this way cannot infect cells and are easy prey for other parts of the immune system that engulf and destroy them.

But if you remember how viruses work, you might have noticed a loophole in these defenses: cells of the body that have been infected with viruses churn out countless copies of the very same invaders the immune system is trying to eliminate. That's where the killer T-cells come in. Guided and stimulated by helper T-cells, killer T-cells learn to recognize and destroy any cells infected by viruses. These dedicated killer cells thus sacrifice the body's own tissues to curtail the spread of infection.

Although this process is incredibly efficient once it gets going, there is obviously a hit-or-miss aspect to it. In addition, the system takes a significant amount of time to get started. That's why several days usually pass between detection of a previously unknown virus and production of enough antibodies to bring the invader under control. During that time, the virus can infect enough cells to cause the familiar symptoms of the flu.

Importantly, some of the B-cells and T-cells stimulated by viral proteins don't mature into active antibody factories or dedicated killer cells right away. Instead, they just hang around while their antibody genes "remember" exactly what those particular viral proteins looked like. These "memory cells" survive in the body for years, even decades, ready to respond rapidly should they ever meet their "chosen" foreign protein. For that reason, a particular strain of the flu can't successfully infect the same person more than once. If it does enter the body a second time, memory cells left over from the first encounter swing swiftly into action, stopping the infection before it ever gets going. Memory cells also enable vaccines to work, for vaccines' function is simple. They "educate" the genes of the immune system by selecting and stimulating those cells genetically programmed to recognize particular pathogens. The result is the benefit of memory cells without the inconvenience, or danger, of infection.

The type of vaccine most widely used against influenza is an injection of mass-produced killed viruses. Vaccine formulation begins with the injection of live virus into thousands of chicken eggs, which are incubated for forty-eight hours. The eggs are then opened, and the virus is extracted and purified. Finally, the virus is treated chemically to kill it by a method that preserves the natural shape of its coat proteins. Each final dose of vaccine is very small, containing scarcely fifteen millionths of a gram of each of three different viruses. When that tiny amount of foreign protein is injected into your arm, your entire immune system is placed on red alert. Assuming that you are inoculated at least a week before you're exposed to the live virus, your B-cells and T-cells will be ready to defend you.

Genetic education of the immune system in this manner can protect you for several years against a number of viruses, including those which cause smallpox, yellow fever, polio, measles, and whooping cough. These viruses, it turns out, are genetic stick-in-the-muds; their genes, and therefore their proteins, don't change much over time. In other words, the protein coat of a smallpox virus that was collected in 1950 has a good chance of closely resembling another smallpox virus collected in 1960. For that reason, a smallpox vaccine produced in 1950 produces memory B- and T-cells that will probably protect against exposure to smallpox ten years later. That, in turn, is why WHO was able to win the war against the variola virus. Once an effective vaccine was developed, it remained effective for years. And once all the people in an area were vaccinated, there was no place left for the virus to reproduce, so it vanished. (Unlike certain other human diseases, smallpox doesn't live in any animal except *Homo sapiens.*)

Genetic conservatism, which seems to run in the variola virus family, is also what enabled Edward Jenner's original vaccine to work. The cowpox virus, it turns out, is a close relative of the much more virulent smallpox virus. Significant stretches of their genetic material—hence substantial portions of their coat proteins—didn't change much after the viruses diverged from their common ancestor. As a result, an immune system educated by the cowpox virus was prepared to do battle against smallpox through a fortuitous case of mistaken identity.

But if that's the case, you might ask, why can't researchers make a vaccine to control the flu as effectively as smallpox? And why is it so hard to come up with a vaccine for AIDS? The answer to these questions summarizes one of the greatest challenges facing medical science, for it explains why the immune system's successful molecular strategy is also its greatest weakness. Precisely because the system focuses its attack on specific foreign proteins, rather than striking out indiscriminately against anything alien to the body, the immune system must recognize a foreign protein before the attack can begin. Therefore, if a parasite such as a virus can either "hide" its proteins from view or change them, even slightly, memory cells produced by a vaccine or previous infection can no longer recognize it. When that happens, the entire process must again start from scratch, so while the immune system is gearing up for a new assault, the parasite has enough time to do serious damage.

That's what happens with influenza virus strains; they change their protein coat frequently enough to keep our immune system perpetually on the defensive. Flu viruses change particularly fast because they don't have any molecular "proofreaders" to check for accuracy as the viral genome is copied inside our cells. As a result, mutations in viral genes occur frequently, and new strains of the virus—each of which is slightly different from the original form—appear all the time. The more different the new strains are from older forms, the more effectively they can avoid the immune system. And the longer a virus can evade the immune system, the better its chances of replicating enough in each host to get spread around in coughs and sneezes.

In effect, therefore, our immune system is a potent form of natural selection that directs the evolution of these viruses. Strains that don't constantly mutate are effectively wiped out. Only those strains which become molecular chameleons are able to defeat, or at least delay, the actions of the immune system, and therefore manage to survive.

Now you can understand why Nancy Cox and her colleagues constantly collect flu strains from around the world and test them to see how they differ from other recent strains. Most of the time, new variants are only slightly different from existing forms and don't cause much trouble. Why? Because many people, having recovered from bouts with similar strains, have memory B- and T-cells left over that recognize those variants and quickly mount an effective defense. Many such strains, therefore, aren't worth putting into vaccines. Other strains are problematic enough to merit inclusion in the annual vaccine, which is recommended for individuals sixty-five and over, for people at high risk for complications of influenza, and for those who hope to avoid the flu in a given year.

But every now and then a *really* different strain appears, one that is a total stranger to the immune system. That can be a dangerous virus. One such variant, the Asian flu, appeared in 1957 and spread like wildfire around the world. Appearing in China in February, it spread through most of Asia and Japan by May and by July it had invaded Europe and the United States. Over the next three months it reached nearly everywhere else. Although antibiotics and improved health care prevented the epidemic from causing as many deaths as the 1918 version, the medical and economic consequences were devastating.

That's why the CDC is always on guard. Its experts play a guessing game in which life and death hang in the balance. Not only must they judge which strains are new enough to trouble the immune system, they must also predict which strains will spread in a manner that will cause serious trouble. Because the flu knows no national boundaries, our CDC works not only with national agencies in other countries, but with global coordinating centers run by the World Health Organization.

Virus samples from all over the world are flown to Atlanta, where they are logged into a computer database that records where the strain came from and whether it caused only isolated cases or seemed to be producing a more serious outbreak. Part of the sample is stored in the freezer, and part is inoculated into chicken eggs and special cell cultures that produce enough of it for researchers to work with. Cox and her colleagues make genetic fingerprints for each new strain and compare them with known viruses. Samples that look as though they might be important are tested to see how efficiently they avoid the immune response. Throughout the entire process, the chosen strains are monitored to make certain that the laboratory cultures haven't mutated too much from the original samples to be useful. In the meantime, in the regions where the viruses originated, physicians and public health officials monitor the effects of each local strain on people in their area. For reasons that still aren't understood, some strains spread quickly, while others don't get around much.

The results of these parallel studies are combined, analyzed, presented, and debated at a meeting of the U.S. Food and Drug Administration (FDA) Vaccine Advisory Group held annually at the end of January. There, infectious disease specialists make decisions about which flu strains are likely to cause the most trouble in the United States the following year. In mid-February, a similar meeting of WHO consultants is held to make final recommendations for the vaccine to be distributed worldwide. Once approval is given to vaccine producers, the rush to manufacture the vaccine is on.

WHILE THE CDC looks ahead to the coming year's dangerous flu strains, Saint Jude's Robert Webster and several colleagues are looking even further ahead, trying to address the nagging problem of the occasional killer flu that takes the immune systems of the entire world's human population by surprise. How do such major variants appear? Where do they come from?

How would flu researchers begin such a search? You can begin to follow in their footsteps by recalling the names of the worst flu strains you've heard about. In addition to the 1957 Asian flu, exceptionally virulent strains have been identified as Hong Kong, Shanghai, and Taiwan flu after the places they first appeared. Because that concentration of localities isn't coincidental, both the CDC and WHO pay particularly close attention to new strains from Asia. But *why* do all these serious variants seem to come from the Far East? The answer, according to one fascinating hypothesis, is hinted at by the name of yet another nasty strain, swine flu.

According to Webster, the search for answers couldn't even start until the molecular revolution had begun. Before that, medical researchers could only guess what was going on. "In 1957," Webster recalls, "when the Asian flu virus appeared, it spread all the way around the globe, because nobody in the world had any antibodies or immunity to it. The WHO said, 'Where did this come from? How did it change so much so quickly?' We just didn't have the tools to understand it at the time."

It was years before the next really big change in the virus appeared. Then the Hong Kong flu of 1968 infected 50 million Americans and left 70,000 dead in little more than six weeks. But this time, the first molecular tools for studying viruses were in place. Webster's colleague Graeme Laver isolated and studied the proteins in the virus spikes. He then examined the spikes of an earlier human flu strain, along with several strains found in ducks. To his surprise, the spikes on the human Hong Kong flu virus looked a lot like the spikes on the duck virus. Indeed, the Hong Kong strain carried six genes from normal human flu strains and two genes from an avian strain. Subsequent studies performed on samples preserved from the 1957 epidemic showed that the Asian flu had carried five genes from human strains and three from avian strains.

Researchers were stunned. They had known for decades that different strains of the influenza virus infect ducks, pigs, and a variety of other animals. These animal infections form what researchers call reservoirs of the virus outside of human populations. Most of the time, however, the majority of those reservoirs pose no threat to us, because the strains that live in these other animals usually can't infect people. "In England," Webster told us, "they did a study on human volunteers. They put the duck virus in humans and it did almost nothing. It didn't induce an immune response. The viruses replicated for a couple of days and then disappeared."

Therefore, the new human flu strains probably weren't simply duck viruses that switched over to humans, as malaria seems to have done. Instead, Webster and several other influenza researchers theorized that human influenza viruses now and then manage to pick up genes from strains that lived in ducks. But if human strains didn't live in ducks, and duck strains couldn't infect humans, how did the genetic exchange take place?

Then Walter Fitch of the University of California at Irvine, along with Christoph Scholtissek of the University of Giessen in Germany, used studies of proteins from many strains of flu viruses to construct an influenza evolutionary tree. That tree split into two main branches: one contained all the strains found in humans, and the other carried all the strains found in birds. That wasn't too surprising. But interestingly enough, depending on just how the data were analyzed, viral strains isolated from pigs either formed their own separate branch on the tree or appeared as twigs scattered between the human and bird branches.

Soon it became clear that pigs could be infected by flu virus strains that lived in both humans and birds. That possibility of dual infection led researchers to a fascinating hypothesis. Influenza viruses carry their genes on eight separate pieces of RNA, so if a human strain and an avian strain happen to infect the same cell, copies of genes from the two strains could accidentally be "packaged" with one another. If this was, in fact, how avian influenza genes entered human strains, pigs could be the missing link, the "mixing vessel" in which the switch takes place.

Circumstantial evidence, including the earliest emergence of virulent flu strains in China, suggests that the chances of such mixing are highest in that country's southern mountains. The region has been heavily populated for centuries, and farmers have long been under pressure to produce as much food as possible in minimal space with minimal input of artificial fertilizer. The result has been the development of a combined aquaculture-agriculture system in which fishes, birds, and pigs are all raised together. Animal feces are used to fertilize fish ponds and farmers' fields, while food wastes uneaten by humans are recycled into animal feeds. In many places, pigs, ducks, and humans literally live under the same roof. Conditions seem ripe for humans and birds to pass their viruses on to pigs, and vice versa.[5]

It's an intriguing and logically appealing hypothesis, but does it accurately describe what happens? In an effort to find the answer, Webster and Guo Yuanji, of the Chinese Institute of Preventive Medicine in Beijing, are conducting a joint study in the mountain town of Yialing (Eee-a-lang), in southern China. "We have twenty families that raise pigs and ducks in the same household," Webster notes, "and each week, all year, we take samples from all the critters in the house. We want to find out how much genetic exchange is really occurring between influenza viruses in these different creatures. These families will certainly get influenza this winter. Will they get the same strains as those circulating just in the rest of the human population or will we see occasional genetic exchange coming across from the duck to the pig to the human? We don't know yet."

Webster and Yuanji hope to use the results of their work to discourage the emergence of new killer flu strains in two ways. First, if they can prove that pigs are indeed the mixing vessels for influenza virus, they may try to persuade the Chinese government to discourage the practice of housing ducks, pigs, and people so closely together. That cultural change would not be easy to achieve, but because it would make cross-species infections less common, the benefits could be enormous. "In a sense," Webster says, "agricultural people are following similar reasoning in the United States. Turkeys [which can also be infected by viral strains from ducks] raised on the range in Minnesota have been moved to ridges and hills, away from river valleys and lakes where the migrating ducks gather."

Second, researchers can look to molecular biology for a source of either more effective vaccines to provide broader immunity to new flu strains or for new drugs to help the body fight viral infections. Both those solutions are still a long way off, but work in these areas, as in much of molecular biology, is proceeding by leaps and bounds.

That research must proceed quickly because we have already eliminated or controlled those few diseases of consequence, like smallpox and polio, which are susceptible to simple vaccination strategies. The world's human population is growing rapidly, and human-caused changes to local environments are increasing in extent and intensity. As we alter local habitats by building cities, erecting dams, or clearing land for farming, we are also altering our relationships with all of earth's life forms, including parasites. In a sense, we are changing the rules of the life-and-death game between host and parasite. Every now and then, one parasite or another takes advantage of such changes, and a new threat to human health emerges or increases in severity.

• • •

It is afternoon in the village of Itaquara, five hours' drive inland from the seaside city of Salvador in northeastern Brazil. The sun beating down from the desert sky is hot. At the edge of town is a shallow, tepid lagoon villagers created some years ago by building a dam across a small stream. The lagoon was meant to be an oasis of calm and pleasure, a place to wash clothes, a place in which to swim and fish, and a place to relax and forget about the harshness of desert life and the ravages of poverty.

There are many children in Itaquara, and today many are out playing, trying to stay cool in an area with few trees or refreshing breezes. Given the heat, it seems odd to see only a few taking advantage of the lagoon. Some splash around in the water, motioning playmates on shore to join them. Others dare one another to balance on logs or planks that span the sluggish, tea-colored stream. Still others stay far from the water, preferring to play hide-and-seek in the shadows of the huts and small houses in which they live.

Nine-year-old Rogerio is particularly careful to steer clear of both the lagoon and the stream, and for good reason. He recently recovered from a disease he picked up from that water, an illness so severe that it almost killed him. "He stayed at home for six days with a high fever," Rogerio's mother recalls. "I gave him a lot of medicine, but he didn't get any better. Then I took him to the hospital. He had to stay there for a month and fifteen days. Every day the fever would be higher. Each hour it would be higher. I prayed to God to cure him."

Tests performed at the hospital confirmed what Rogerio's doctors suspected. The boy had contracted a near-fatal case of schistosomiasis, a tropical disease caused by a parasitic worm that infects somewhere between 200 and 300 million people around the world—between 5 and 10 percent of the world's population. Luckily for Rogerio, his mother got him to the hospital in time, and a course of drug therapy killed the worms inside his body. Had he not received that treatment, his name would have been added to the estimated 2 to 3 million annual fatalities from "schisto," as doctors and researchers call it. WHO estimates that nearly one billion people are at risk for the disease, making it second only to malaria in global devastation.

Ironically, schistosomiasis isn't native to Brazil, but then again, neither were the ancestors of many people in Rogerio's community. Most of Itaquara's inhabitants, who are either black or very dark skinned, are descendants of Africans the Portuguese brought over as slaves to work sugar cane fields and cacao plantations during the sixteenth, seventeenth, and eighteenth centuries. With those slaves and the European colonists who brought them came a host of diseases, including typhoid, smallpox, and yellow fever, whose grim history is recorded by tombstones across the region. Many of those diseases are more or less under control in Brazil today. Schisto, unfortunately, is not.

The worm that causes schistosomiasis employs a successful strategy that is fascinating in a gruesome sort of way; its life cycle is complex and unpleasant enough to serve as a model for a film like *Aliens* or *Invasion of the Body Snatchers*. In order for the worms to thrive, four crucial elements must be in the right places at the right times: human hosts, human wastes, quiet water, and one of several species of snails in which the parasites live for part of their lives.

The story begins as parasite eggs hatch in water, where larvae disperse in search of snails to serve as their first hosts. After growing, feeding, and reproducing within the snails for a time, the parasites wait for the propitious moment to depart. Then, stimulated by sunlight, the worms leave the snail and swim through the water, searching for exposed human flesh. When they find the legs of a woman washing clothes or the arms of a child swimming, they burrow into the skin, enter the circulatory system, and travel with the blood to the vessels of the intestine. There they mature, mate, and begin laying up to 300 eggs each day, many of which break through into the intestine itself and pass out of the body with the feces. If those wastes are carried into snail-infested bodies of water, the eggs can hatch and the cycle begins again.

Unfortunately, the immune system seems incapable of attacking and destroying the adult worms. "It's phenomenally interesting," notes Don Harn, a molecular biologist at Harvard's School of Public Health, "because you've got this foreign body inside of yours, and your immune system should be attacking it and killing it, but it doesn't. The adult worms live right in your blood, right where all the immune system parts are, yet the parasites may live for as long as twenty years."

How do the worms, surrounded by the host's immune system, survive? They conceal themselves from molecular surveillance with tricks that enable them to blend in with their surroundings. Like human guerrillas who cover themselves with tree branches and paint their clothing to match their environment, the worms cover themselves with host proteins and cells and produce different proteins that the body cannot recognize as "other." These tactics combine to conceal the foreign proteins of the worm from the receptors carried by macrophages, B-cells, and T-cells. By default, therefore, the body treats the worm as self rather than other.

If you can imagine hundreds or thousands of such camouflaged worms living in your blood vessels, you probably understand that they can make life unpleasant, particularly for a child like Rogerio. The worms eat red blood cells, so a heavy worm infestation causes anemia, which can damage the developing brain. Ironically, though, the most serious symptoms are caused not by the worms themselves, but by the body's misguided efforts to fight them.

In a strange twist of fate, although the immune system fails to recognize adult worms, it does mount a strong response against their eggs, a response that harms the host in the process. Although many eggs leave the host's body through the intestine, many others are carried around by the circulatory system and become lodged in the liver. Left alone, the stranded eggs would slowly deteriorate and cause few problems. But the immune system recognizes them and mounts a furious assault, which slowly but steadily blocks blood flow to the liver. At the same time, the spleen, a major production site for the cells of the immune system, becomes overworked and stops functioning properly.

Unfortunately, the manner in which schistosomiasis finds its victims and the long, slow process by which it causes disease make it tough to control in rural areas. Itaquara's lagoon and streams, filled with snails, are surrounded by houses without running water. Sanitary facilities—if one can call them that—are simple huts built out over the water; raw human sewage loaded with schistosome eggs flows into the lagoon constantly.

Public health officials first tried to eradicate the snails. "They failed," Harn explains. "Miserably. What happened is that they killed all the snails, not to mention beaucoup wildlife. The drugs were so toxic that they knocked off the plants, the fish, everything. And the water couldn't be drunk by anybody. Then, when the next big flood came through, *voosh!* It brought in all new snails from the next site, so it did no good. They might have gotten rid of snails for six months, maybe a year, but bingo, they are back."

Then came a massive public education program, an attempt to explain to people how the disease was spread. That didn't work either. Established behavior patterns are hard to change, especially among poor people with few alternatives. The calm waters of the lagoon are enticing to children like Rogerio, who grow up in tiny, unfurnished homes without toys in a village without playgrounds or parks. And afternoon gatherings in the stream to clean pots and wash laundry are social traditions among village women.

On top of that, it's hard to change people's behavior because the parasites take so long to cause serious illness. "You tell people," Harn says with a sigh. "You say, 'You're going to get schisto and in ten to fifteen years your spleen is going to blow up and your liver will dysfunction and you'll die.' But most people—in Brazil, or anywhere else for that matter—don't relate to that kind of advice. It's not an immediate impact. It's not like malaria, from which you die tomorrow. So they continue to go into the water. They see the old people with scars, and their spleens are missing and their bellies are swollen, but it's a remote thing—ten years down the road."

Part of the problem is that even entering the water once or twice courts serious infection. Rogerio's story is a case in point. "One day we went out bird hunting," he responds when asked how he got the disease, "and then a boy said like this: 'Man, I'm so hot! Let's go to the river and cool off. Let's go!' Then he took my clothes off and pushed me in the water. So I stayed in the water. Then we went bird hunting again. Then I got sick."

Rogerio's mother made an even smaller mistake. "I remember once when I went in just to get the children out," she recalls, shaking her head, "and I caught it. I've never taken baths there, I've never washed clothes or dishes there, and they found the worm in me. I suppose that it was when I went in the water to get the children that it went inside of me."

What is worse, the disease is spreading, and spreading fast. The reason, notes Harn, is obvious. "It is flat-out due to the spread of agriculture. They are irrigating, and with irrigation comes the snails. That obviously is not going to diminish. It is going to continue to increase. People are becoming more agricultural."

All these factors cloud the future of northeastern Brazil. Harn and his colleagues agree that education and sanitation help, and that drug treatment of severely infected people is an invaluable aid. But individuals who expose themselves to contaminated water can be reinfected within weeks of treatment. The only way to keep so many people from getting sick is to somehow educate their immune system to handle the worms more advantageously.

Ideally, Harn and his coworkers, Evan Secor of Harvard and Mittermayer Reis of Brazil's Findacao Oswaldo Cruz, would like to produce two changes in the immune response to schistosomiasis: a more rapid assault on invading larval worms and a truce with parasite eggs in the liver. But is it possible to tinker with the immune response that way? Harn notes realistically that despite all the successful vaccines against viruses and bacteria, there has not been a vaccine that works against larger parasites in humans.

However, challenges inspire this team. In the rear of Itaquara's small hospital, they've set up a clinic to examine and treat villagers while studying their responses to parasite infection. They have discovered what appear to be several genetically based differences in individual villagers' immune responses to worm attack. Some people—Rogerio and his mother, for instance—are highly susceptible, both to infection and to serious disease. But some of their neighbors are much less likely to become infected. Still other villagers, while they do become infected, tolerate the parasites without getting very sick.

These differences are striking, especially when compared with the condition of North Americans carrying even a few worms in their bodies. "If you were infected with an almost undetectably small number of worms," Harn warns, "you would be very, very sick. You would think you were going to die. You could get that sick from just enough worms to give you twenty-four schistosome eggs per gram of your feces. A patient with that level of infection in Brazil may not have any symptoms, would not be recognized as ill, and therefore would not be treated. We have patients in Brazil who carry many more worms than that yet have no serious symptoms—at least not so far!"

Findings like these suggest to researchers that human immune systems are capable of responding to schistosomiasis more productively and less self-destructively. The trick is not only to find out why that is sometimes so, but to learn how to force the system to respond that way all the time.

In seeking such an alternative treatment, Harn, Secor, and Reis must look to the villagers' genes—the particular stretches of DNA that control their immune systems. Their pursuit forces the team to work long hours under difficult conditions. They draw blood from villagers and rush back to Salvador in a car filled with Styrofoam coolers. There they proceed as quickly as possible, isolating the cells of the immune system, challenging them with several proteins isolated from worms, and observing their response.

Laboratory facilities in Salvador aren't quite what they are back home. Several types of immune system cells grow best, for example, in culture dishes whose air is enriched with extra carbon dioxide. In Boston, the researchers have a tank filled with carbon dioxide; in Salvador they don't. Says Evan Secor with a smile, "We found that you can simply add an Alka-Seltzer tablet to the water. It generates the gas we want. Keeps the cells very happy."

The makeshift appearance of their field equipment, however, belies the sophistication of their research. They are studying not just one particular group of immune cells, but rather the interactions among several different cell types, all of which communicate with one another to coordinate the body's response against schistosomes.

The differences in their Brazilian patients' response to schistosome, it appears, can be traced to the activity of the immune system's "central command," helper T-cells which, like B-cells, carry receptor molecules that latch on to foreign proteins. Millions of different T-cell receptors are produced by the same sort of gene reshuffling that goes on in B-cells.[6] T-cells, too, are stimulated to divide when their receptors match up with the right foreign proteins.

The T-cells involved in the attack on schistosomes, however, don't work on their own, and many of them don't do any actual fighting. Instead, helper T-cells act like commanding officers in a human military organization; they coordinate and direct the attack by engaging in constant dialogue with other immune system cells. T-cells learn from the system's advance guard which invaders are present in the body. In return, they tell other types of cells which of those potential targets to strike.

Because adult schistosomes' molecular camouflage prevents the immune system from "seeing" them, the body normally can't do much about worms living in the intestines. Yet the researchers found evidence that certain antibodies and certain types of T-cells could attack the worms, if only they could see them. The drug used to cure Rogerio, in fact, apparently works by somehow stripping the worms of their camouflage and exposing them to the body's defenses. There is other evidence, too, that at least some people who've been purged of severe schistosome infestations, like those who've recovered from malaria, seem to be resistant to further infection.

Harn and his colleagues take this as encouraging evidence that they can develop a protective vaccine to teach T-cells to attack invading worms *before* they put their camouflage in place. Even if this stronger, faster response can't eliminate 100 percent of the invading worms, it would still be enormously useful. "The World Health Organization has decided," Harn notes, "that if a vaccine was only 40 to 60 percent effective—if it could reduce the number of parasites in every person by 40 to 60 percent—it would eliminate most of the disease. You would have an infection, but you might not have the painful and dangerous symptoms of the disease." That, of course, could allow millions of people to live much better lives.

But the search for a protective vaccine to attack adult worms is only half the story. The most serious symptoms of the disease are caused by the immune system's misguided attempt to attack worm eggs in the liver. If that attack could be turned off, instead of being encouraged, serious symptoms would be significantly reduced. To address that possibility, researchers have had to determine exactly what aspect of the immune response in the liver causes the damage.

It turns out that because schistosome eggs aren't camouflaged as well as adult worms are, macrophages can find, recognize, and attach to them easily. But macrophages can't launch a full-fledged attack on their own; instead, they send out a molecular message that translates more or less as "We've found an unusual protein here! Should we do anything about it, or do we let it pass?" Sooner or later, helper T-cells happen by and recognize that alert. Stimulated by its message, they "investigate" the egg proteins. By chance, some T-cells carry receptors that recognize one or another of the foreign proteins. When they do, those particular T-cells send another message to the macrophages: "We don't want this around. Destroy it!"

After that, macrophages and T-cells envelop the eggs and cover them with a sort of isolation chamber composed of living immune system cells. But the immune system works overtime, the isolation chambers build up, and their presence in the liver causes trouble. Scar tissue forms, local blood vessels are blocked, and parts of the liver begin to die.

Armed with this knowledge, researchers are trying to develop a completely different vaccine aimed at modifying the immune response to the eggs. The sort of treatments they are seeking, however, would have precisely the opposite effect of most vaccines. Instead of *encouraging* the body's attack, such a vaccine would *suppress* it, somehow altering the immune system's genetic program to induce macrophages and T-cells to ignore the foreign proteins carried by the eggs. The difficult part is that this strategy must be extremely specific; Harn and his group must manipulate the molecular messengers and genes of the immune system so that they turn off only the specific T-cells that direct the attack against schistosome eggs. General suppression of the immune system is far too dangerous, as AIDS has taught us.

Such a strategy might work, at least in theory, if a method could somehow be found to persuade the immune system to treat the parasite eggs as self rather than other. This type of "tolerance vaccine," although it wouldn't eliminate adult worms, could restore normal function to a patient's liver and spleen. That, for highly susceptible people like Rogerio and his family, could make a world of difference.

Harn, Secor, and Reis, like many researchers studying tropical diseases, are at once breaking new ground and working overtime to make up for basic studies that were not performed earlier. Many of these diseases have traditionally been ignored or slighted by European and American researchers for geopolitical reasons. They didn't directly affect "us," after all. What they did do was make very large numbers of very poor people very sick, predominantly in Third World countries.

The unifying principles of molecular biology have proved that molecular studies of host-parasite interactions inevitably lead down converging paths to similar cells and genes. So it is interesting but not surprising to find connections between the efforts of Harn and his colleagues to devise a tolerance vaccine for schistosomiasis and the struggles of patients in North America who never have been—and hopefully never will be—bothered by that parasitic worm. These people are, however, plagued by an incorrectly directed and overactive immune system that clinical researchers are trying to redirect.

• • •

HALF A WORLD away from Itaquara, a meeting room in Lexington, Massachusetts, fills slowly but steadily with patients gathering to hear a lecture on T-cells. An uninitiated observer might be hard pressed to imagine what all these people have in common. Some have lost the sight in one eye. Others experience numbness, tingling, or weakness in their arms or legs. Some have had such trouble walking that they spend much of their lives in wheelchairs.

Though their symptoms vary, most of these people suffer from multiple sclerosis—MS. Multiple sclerosis is a severe autoimmune disease, a condition that occurs when the body's powerful cellular and molecular arsenal turns against some of the very tissues it is meant to protect. In rheumatoid arthritis, the attack is aimed at joints between bones. In juvenile-onset diabetes, a misguided immune system strikes insulin-producing cells in the pancreas. And in MS, the body's weapons assault the central nervous system, damaging the brain and spinal cord and causing weakness, blindness, and even paralysis.

In all these diseases, renegade T-cells recognize and attack targets they're supposed to ignore, proteins they should recognize as self and leave alone. Researchers struggling to combat MS and other autoimmune conditions all share a common goal: a safe and dependable way to blunt that attack. If this approach sounds familiar, it should. The type of treatment sought by autoimmune researchers is similar to that of Harn's hoped-for second vaccine, a procedure that uses molecular techniques to identify and either eliminate or shut down the rogue T-cells that damage the body.

On this particular evening at the National Multiple Sclerosis Society gathering in Lexington, the keynote speaker is David Hafler, codirector of the MS program at the Brigham and Women's Hospital in Boston. His work offers a new glimmer of hope for treatment of autoimmune disease. Keenly aware of both his power and responsibility as a physician, Hafler discusses discoveries and hypotheses carefully, qualifying them with many cautious disclaimers. He is the first to admit that his laboratory work is enormously exciting, but quickly emphasizes that many hurdles and potential problems stand between a culture dish in an incubator and a safe, effective treatment for MS patients.

Hafler begins by explaining that multiple sclerosis means "multiple scars," referring to the tough, damaged regions in the brains of MS patients. On slides generated by a computer program in his laboratory, he shows three-dimensional reconstructions of a patient's brain, revealing the scarred areas. He explains that he and his coworkers have isolated and cultured a distinctive type of T-cell from such areas in brains recovered during autopsy of MS patients. They have also identified the same type of T-cells in samples of spinal fluid taken from some of the very people in that room.

Hafler has confirmed that those specific T-cells are doing something wrong; they are attacking the protein myelin, which forms a protective sheath around nerve cells. When myelin sheaths are removed, and when neurons inside are disturbed by the immune response around them, those nerve cells begin to degenerate. Ultimately they die.

Where do such T-cells come from? Why have they run amok? Is there any way to kill them or shut them down? In tackling these problems, Hafler—like Harn and everyone else trying to manipulate the immune system—has dedicated (some might say condemned) himself to wrestling with one of the body's most complex and Byzantine networks. Their work strikes close to the core of the immune system: its ability to distinguish self from nonself.

Normally, the immune system recognizes and respects the critical difference between "self" proteins that belong in the body and "other" proteins that don't. Largely during fetal life and infancy, the immune system's cellular command personnel, the helper T-cells, are rigorously instructed that molecules found in the body's tissues are to be treated as self. That molecular definition of self, when it works properly, is incredibly specific; it can distinguish your cells from those of anyone else in the world, except those of an identical twin. This much is clear. But precisely how T-cell education takes place and just why it sometimes fails are matters of intense and often acrimonious debate.

There are some parts of the story about which nearly everyone agrees. T-cell receptors, as noted earlier, are produced by the same kind of random gene shuffling that produces B-cell receptors. As a result of that genetic rearrangement, some resulting T-cell receptors inevitably, just by chance, recognize proteins such as myelin that belong in the body. It could be dangerous to have hordes of those T-cells wandering around the body, because if their receptors recognize self proteins, they can direct attacks against any body cells that contain those proteins. But that danger is minimized in what might seem to be a wasteful manner; early in life, many potentially self-reactive T-cells are destroyed.

How does the body single out dangerous, self-reactive T-cells and get rid of them? The key to that process seems to be the thymus gland, a small but vital organ that sits behind the breastbone. The thymus somehow manages to display on its cell surfaces a large number of proteins that belong in the body. Its display seems to act as a sort of molecular catalogue of most proteins that are part of self, namely, any protein displayed there should not be attacked.

Very early in its life, a T-cell makes an obligatory passage through the thymus. As it does so, it effectively browses through the molecular catalogue, searching for proteins on display that match its receptors. But this is not the kind of catalogue shopping you would like to do, for the items on the shelves, as it were, are attached to molecular booby traps. Any T-cells whose receptors attach to proteins presented in the thymus are summarily destroyed. Only those which pass through without recognizing anything survive, mature, and circulate through the body.

It was once thought that this process was foolproof, that *all* the body's proteins were displayed there, and that *all* T-cells able to recognize self proteins were destroyed. But that, like many other early hypotheses about the immune system, no longer appears to be true. While *most* of the potentially dangerous self-reactive cells are removed, *some* are not. Many immunologists feel that only a few slip through, mainly by accident. Hafler feels otherwise.

"We know that's not the case," he asserts back in his laboratory. He leads us to an incubator, removes a culture dish full of T-cells, and places it under a microscope. "Take a look at these," he exclaims. "These cells go crazy when they see myelin basic protein. They recognize brain tissue." That in itself is no longer surprising; Hafler has, after all, isolated myelin-reactive cells from MS patients, but these particular cells aren't from an MS patient. They are descended from cells that originated in Hafler's own blood.

Is Hafler worried about getting MS because some of his cells recognize myelin? No. He is certain that all of us carry those cells, despite the fact that most of us never suffer from multiple sclerosis. "We've established over the last three to four years," he asserts, "that T-cells in the blood which recognize brain tissue are the same in MS patients as they are in everyone else. They're there. They have the same T-cell receptor. The key is that only in MS patients do they become active."

The reasons why different self-reactive T-cells may become active in different people are the subject of the debate and acrimony mentioned earlier. Some immunologists feel that those potentially rogue T-cells are actively turned off. Others feel that they are naturally off and must be turned on or activated in some way before they swing into action. Hafler's group, for example, feels that myelin-recognizing T-cells in MS patients are activated either by certain types of viral attack or by inflammation in the nervous system. Still others feel that those T-cells are always potentially on but are normally suppressed by other immune system cells via unexplained molecular dialogues.

The controversies and conflicting hypotheses in this area are as bewildering to outsiders as the immune system itself is to researchers. Which ideas are correct? The interesting point to consider here is that all of them may be partially right, none of them may have the whole story, yet each of them may lead to potential treatments.

Consider the old Indian fable about four blind men trying to describe an elephant. Because none of them could see the beast, each used his sense of touch to investigate the first part he came across. One, examining the trunk, exclaimed, "The elephant is a thick, leathery snake that coils and uncoils." The second, who happened on the tusks, cried, "No, the elephant is smooth and hard and tapers to a pointed tip." The third, who discovered the legs, responded, "Nonsense! The elephant is leathery, but it is stiff and upright, like a tree trunk." And the fourth, who grabbed the tail, replied, "You're all wrong. The elephant is slender, ropelike, and twitches this way and that."

The similarity between the elephant researchers and molecular immunologists is greater than you—and some scientists—might think. Researchers, though they can see their cells under a microscope, *can't* see the difference between all the various types of T-cells we've talked about, so they are effectively blind in that respect. They can't actually watch the immune system operating in its entirety because there are just too many activities taking place all at once. Instead, they must use specific molecular tools to ask precise questions. As a result, researchers who concentrate on different parts of the immune system come up with different descriptions of what's happening. Just like the blind men, each of them can at once be entirely correct about a piece of the system and mistaken about the appearance of the whole beast.

However, partial answers can be very useful. Each immunologist who feels that self-reactive T-cells are turned on, turned off, or suppressed can suggest a therapy with a good chance of working. For example, one of Hafler's most promising experiments takes advantage of the fact that, somehow, certain parts of the immune system can suppress other parts. With colleague Howard Weiner, he is investigating an entirely new kind of vaccine that has extraordinary potential, not only for MS and other autoimmune diseases, but perhaps schistosomiasis as well. That vaccine wouldn't be injected into patients—it would be fed to them.

By way of explanation, Hafler and Weiner show us laboratory mice in various stages of paralysis caused by a mouse version of MS. Some are sluggish, while others seem unable to move at all. But in a few cages, mice that were nearly paralyzed several days earlier scamper around actively. Those mice, Weiner explains, were fed a special oral vaccine the team designed to suppress rogue T-cells that attack myelin—a vaccine that contains a rodent version of the myelin protein. But how could that sort of vaccine work? How could swallowing myelin turn off myelin-reactive T-cells?

"You've got to eat," Hafler points out. "Everything you ingest is foreign to the body. If you developed an immune response to everything you took in by mouth, you'd be in very deep trouble." He continues to explain that there is a large and only partially understood branch of the immune system located in the lower digestive tract where digested food, which almost always contains bits and pieces of foreign proteins, is absorbed into the body. Why don't those foreign proteins elicit an immune response? Because they are detected by this little-known branch of the immune system, which acts to prevent the rest of that system from attacking those particular proteins.

Hafler and Weiner hypothesized that this system might work by suppressing the T-cells which would otherwise react to the proteins in food. To test that idea, they fed myelin to mice that were suffering from experimentally induced MS. To their delight, the oral tolerance vaccine, as they call it, actually worked! The myelin-reactive T-cells in the experimental mice were suppressed, though no one knows exactly how. "If you consume this protein," Hafler notes, "it turns off the T-cells for about three months. It's like a negative vaccination. You'd probably take it every day."

The team is conducting clinical trials of the procedure on a selected group of human patients, although Hafler urges extreme restraint and caution in interpreting the earliest results. "As physicians who care for patients," he reminds us, "we're not interested in everybody jumping all over this right now, because it still might be a false hope. If it works, it works, and it will be fine. We have enough resources at our disposal to proceed as quickly as we need to. I would much prefer to do the work in relative peace and quiet."

That peace and quiet may be harder for Hafler to find in the near future. Parallel studies are providing preliminary indications that the procedure might also work to control other autoimmune diseases. The potential of oral tolerance vaccines is staggering. If, for example, Don Harn and his team can determine the proteins responsible for eliciting the attack against schistosome eggs, an oral tolerance vaccine made from those proteins might conceivably do just what is necessary to save their patients' livers—and their lives.

• • •

THE DISEASE APPEARED in Africa, where our species evolved and where humans still live alongside our closest primate relatives. That alone isn't surprising; virulent new parasites crop up in Africa with alarming frequency. We just don't hear about them because they either kill off isolated groups of people and disappear or become less virulent and peter out before leaving the hinterlands. But this particular disease wasn't just new to science. It was also new in its mode of action; it infiltrated the immune system, lived inside it, and attacked its central command, leaving its troops leaderless and powerless to act.

It spread through Africa, remaining invisible for years. It rarely killed people by itself; rather, it allowed them to die at the hands of other pathogens, especially tuberculosis. It spread slowly to Europe and Haiti, then to the United States, Latin America, and Asia. The first cases—a handful in Europe and one or two in the United States in the 1960s and 1970s—were mysteries that left physicians wringing their hands helplessly and wondering what could be going on. For no apparent reason, people were being killed by normally harmless microorganisms.

Finally, in the late 1970s, this new disease found its way into three groups of people whose behaviors transmitted it with deadly efficiency: the gay community, where sexual freedom was part of self-expression; prostitutes, for whom sex with multiple partners was their profession; and intravenous drug users, for whom sharing needles was a common bond. At the same time, the ailment spread quietly but rapidly among heterosexually active individuals throughout central Africa.

It was first described as a distinct condition in the United States in the early 1980s. Physicians in California and New York began reporting patients with unusual infections and purple skin tumors so rare that most cancer specialists had never seen them. The cancers were identified as Kaposi's sarcoma, normally found only in patients whose immune systems were malfunctioning, either because of advanced age or because of intentional suppression with drugs such as cyclosporine. Most of these people died of pneumonia caused by a fungus, *Pneumocystis carinii,* which had previously been seen only in immunosuppressed individuals.

A few prescient physicians soon recognized that the ailments they treated were only by-products of an invisible infection. Kaposi's and *Pneumocystis* pneumonia were opportunistic infections, successful only because the patient's immune system was impaired for some other reason. Ultimately, the condition was called AIDS, for *a*cquired *i*mmune *d*eficiency *s*yndrome.

Because AIDS was first reported among gay men, many physicians, rather than looking for an infectious agent, assumed that something about these men's lifestyles had destroyed their immune systems. But a few epidemiologists suggested by the fall of 1981 that AIDS was caused by an infectious agent spread through blood and sexual intercourse. By 1982, the first cases were reported among infants born to infected mothers, as well as among hemophiliacs and surgical patients who had received transfusions of blood or blood products.

In 1983, Luc Montagnier at the Institut Louis Pasteur in France—who would be engaged for years in a rancorous dispute over priority with Robert Gallo at the National Institutes of Health in the United States—identified a virus that has since become known as HIV, for *h*uman *i*mmunodeficiency *v*irus. HIV, it turned out, was a retrovirus, which uses the enzyme reverse transcriptase to make DNA copies of its RNA genes.

Our encounter with HIV provided a dramatic demonstration of molecular biology's power to aid medical science. Had this sinister virus appeared twenty years earlier than it did, researchers wouldn't have been able to identify it, because the molecular tools necessary to observe and describe its actions didn't exist. As it happened, a remarkably detailed description of the virus's life cycle has been teased from sick and dying patients at an unprecedented rate. We now know that HIV confronts us with the challenge of defeating the "ultimate parasite"; its combination of molecular stealth and evolutionary cunning makes it not only deadly but incredibly difficult to attack.

For a start, the virus targets the body's T-cells with deadly skill and efficiency. Certain parts of HIV's protein coat are shaped so that they impersonate a vital molecule which normally carries messages from cell to cell within the immune system. That maneuver enables HIV to act like a subversive who has stumbled on a vital password to the enemy's command center. Essentially the protein masquerade enables HIV to grab on to helper T-cells receptors intended to recognize the body's own molecular messenger. Once that attachment is made, the T-cell treats the virus like an invited guest, opens its doors, as it were, and takes the invader inside.

After entering the cell HIV does not immediately use its genetic material to make copies of itself. Instead, it uses reverse transcriptase to make a DNA copy of its RNA genome. The viral DNA then insinuates itself into the host cell's chromosomes so seamlessly that it becomes part of the T-cell genome for the rest of the life of that cell. Henceforth, the host cell treats those genes as its own, maintaining them and duplicating them in perpetuity. That is the most intractable aspect of HIV infection: once a virus has infiltrated itself into the genome, we know of no way to remove it. At present, a cure for AIDS, at least in the traditional sense of eliminating the invader, is scientifically impossible.

The virus doesn't necessarily begin its lethal work immediately, however. After viral genes have entered the host chromosomes, they may remain dormant for months or even years. But when stimulated in unknown ways, viral DNA begins to direct the synthesis of viral proteins. Sometimes it works slowly, producing new copies of the virus without killing the host cell. At other times the virus replicates rapidly, causing the death of the T-cell that carries it. When that happens, the infected person is in serious trouble. For helper T-cells, as you've seen earlier, are the immune system's central command. Once they are incapacitated, macrophages, B-cells, and killer T-cells are rudderless; deprived of guidance from stricken helper T's, they are unable to respond effectively, either to HIV or to other infections. In effect, this virus leaves us with no immune system at all.

What's more, HIV also infects several other cells of the immune system, and it can hide inside those cells for sixteen years or longer before causing disease. Throughout that period, the virus is present only as a string of DNA incorporated into the host cell genome. Because the immune system can recognize the presence of infection only when viral proteins are being made, the virus is absolutely invisible to immune surveillance as long as it remains quiescent.

Furthermore, in order to reach the wandering T-cells, HIV manages to subvert the portion of the immune system located in the gut and urogenital tract. Once introduced to any of the sexual organs or the lower intestinal tract as a result of intercourse with an infected partner, the virus uses its mastery of the immune system's molecular language to hitch a ride aboard cells that normally carry foreign proteins to other parts of the immune system. Those cellular transporters ferry the enemy directly to its targets.

But the crowning glory of this virus—from its own perspective, of course—is its mode of transmission. If an evil genius had wanted to create a pathogen whose mode of getting from one person to another would stymie twentieth-century public health efforts, he couldn't have done any better. Even though our culture uses sex to sell everything from toothpaste to automobiles, we can't talk comfortably about it in public. Although reason is now beginning to penetrate some government offices and television networks, neither entity was willing to advocate sexual safety measures besides abstinence for more than a decade.

Michael Merson, director of WHO's global program on AIDS, estimates that by mid-1993, more than 13 million young people and adults and more than 1 million newborns had been infected with the virus worldwide. William Haseltine of the Dana Farber Cancer Institute in Boston emphasizes the gravity of the situation. "By the end of the decade," he points out, "all of these people will have died or become extremely ill. There is now no population free of this disease. The true horror of what is happening is just now dawning. We are in the midst of the greatest epidemic of this century, an epidemic which may become the most serious disease in recorded history."[7]

In the face of those figures, our cultural incapacity to halt the virus's spread, and our scientific inability to destroy it once it has entered the genome, place the burden squarely on the development of vaccines and treatments. Yet the attributes of HIV make vaccines difficult to come by. We've already explained that the virus is invisible to the immune system as long as it hides within the genome. Even when actively replicating, HIV adopts every conceivable strategy to avoid our defenses. Much of its protein coat is covered by sugars, a trick that prevents immune receptors and antibodies from recognizing it. The few parts of the protein coat that are not hidden change constantly; reverse transcriptase makes mistakes, so HIV genes mutate rapidly, even during the course of infection of a single individual. Thus, the question is not whether a current vaccine will work for the following year's variant, as in the case of influenza, but whether this week's vaccine will work next month in the same patient!

On top of all this, HIV's ability to infiltrate different cell types makes it impossible for researchers to study the true course of infection in a test tube. Traditional animal models used to study human disease are, unfortunately, of limited use. Chimps and gibbons, for example, can be infected with HIV but don't get AIDS; despite the fact that those animals are close to us genetically, HIV's actions are tightly tied to the human immune system. Mice—favorite models for the study of other diseases, including schistosomiasis and MS—don't become infected by HIV at all. Why? Because mouse T-cell receptors are different enough from ours that the virus cannot latch on to them.

THOSE DIFFICULTIES posed more than a research problem to Mike McCune who, like many other San Francisco physicians during the early 1980s, saw patient after patient succumb to AIDS. "I once asked a patient with advanced *Pneumocystis* pneumonia whether he wanted to be revived if his heart stopped beating," McCune recalls. "I had nothing to offer him." That experience radically changed what it meant to be a doctor.

"Before AIDS," McCune explains, "people in the hospital didn't really talk about death as inevitable. We always had anecdotes about the miraculous procedures we performed—for instance, keeping somebody with severe liver disease alive. But there was never an anecdote about anyone recovering from AIDS."

McCune wanted desperately to make a difference in the struggle against AIDS. Though his work was going as well as could be expected, he knew that watching T-cells in culture dishes couldn't tell him what happened to the immune system of a person infected with HIV. He needed an animal model, one complex enough to truly mimic reality. "It had to have within it all the target organs that HIV could infect in a human," McCune observes. "That includes all the different organs in which helper T-cells can arise, including bone marrow, liver, and thymus, and all the other places HIV shows up—lymph nodes, skin, gut, lung, and the lining of the entire urogenital tract. All those things I just named are the interface between the inside of you and the outside world." But how could he produce a model that complex?

While McCune was driving home late one night, something clicked in his mind. It wasn't the kind of idea he was likely to come up with in the laboratory. It wasn't—as Nobel laureate François Jacob would have called it—day science. Day science, Jacob wrote, employs reasoning that meshes like gears. It solves small puzzles through established experimental methods. Night science is the stuff of intuitive leaps and wanderings—it happens when "thought proceeds along sinuous paths, tortuous streets . . . in quest of a sign, of a wink, of an unforeseen connection." It was night-science thinking that propelled McCune to make a giant leap of faith. He decided that he would transplant a human immune system into a mouse to give him what he needed: an animal model that could be infected with HIV and used to test drugs and vaccines.

McCune is disarmingly modest about that accomplishment. "Not all that difficult," he demurs. But as recently as 1988, when he presented his plan to the board of advisers he assembled to help him, David Baltimore, who shared the Nobel Prize for discovering retroviruses, responded that it didn't seem feasible. It was a great *gedenken* experiment, a thought experiment that could be done beautifully—but only in the mind.

However, the techniques of molecular biology improved enormously in a few short years. Today McCune dons a surgical mask, cap, gown, and gloves, opens a tightly sealed door, and enters a sterile room filled with cages of mice. The animals don't appear unusual; small and white with pink nose and ears, they look like thousands of other laboratory animals. But these aren't normal mice. They're not even entirely mice. The immune systems operating in their bodies come from a different species—*Homo sapiens*. Therefore, these mice can get AIDS.

Describing procedures that took two full years, McCune explains that his team started with mutant mice born without an immune system. This strain, SCID—for *s*evere *c*ombined *i*mmuno*d*eficiency—is completely defenseless against parasites and can be maintained only in sterile quarters. McCune realized that because they had no immune response of their own, these mice would not reject the cells of a foreign immune system transplanted into their bodies.

That was half the problem solved, but McCune knew that he couldn't just inject mice with immune cells from adult humans. Those cells, after all, had been "educated" in human bodies; they would treat mouse tissues as "foreign" and destroy them. The way out of that predicament came from an unusual source: human tissue derived from elective abortions. Fetuses, it turns out, carry tissues that have the ability to make an immune system, but have not yet done so. Some of those tissues contain the stem cells that give rise to all the different types of T-cells and B-cells, while others, like the thymus, are vital in the process of "training" T-cells to distinguish self from nonself. In the fetus, those cells, not having been programmed, are still largely naive.

Going out on a limb, McCune's team transplanted human fetal tissue from thymus, liver, bone marrow, and lymph nodes into SCID mice. After a tense observation period, it became clear that the procedure accomplished just what McCune hoped it would: the human fetal thymus, maturing in a mouse, trains developing T-cells recognize mouse tissues as self. The result was a mouse with a human immune system. Ultimately, these mice were dubbed SCID-hu, the "hu" standing for human.

The final test came when McCune injected SCID-hu mice with HIV. As he hoped, the virus spread to all corners of the immune system, just as it did in humans. The *gedenken* experiment had become reality: it produced an animal model on which to test AIDS drugs and vaccines.

CAN EVEN THESE extraordinary experimental models make it possible to develop a vaccine against AIDS? Many immunologists, asked for opinions in confidence, express both doubt and determination. All researchers, privately and publicly, agree that if such a vaccine can be developed, it will require every bit of our knowledge of immune system function—and more.

The most fundamental obstacle is that most vaccines we've developed thus far protect against *disease,* but not infection. "Vaccines against measles and rubella," Alfred Saah of Johns Hopkins reminds us, "don't prevent you from *catching* those infections. They keep you from getting sick when you get infected. With HIV, if you get infected, the DNA becomes integrated in your cells. So if you don't prevent infection, you may not have done very much."

Another disheartening observation is that the whole principle of vaccination, as we described it earlier, relies on the ability of the immune system to guard people who've recovered from a disease against a second assault. Unfortunately, because AIDS is a disease from which people do not recover, we have no evidence that immunity is, in fact, possible. "If natural infection itself doesn't result in any immunity," Saah notes, "what hope do you have of generating artificial immunity? It's a particularly tough problem." "This doesn't mean that a vaccine cannot be designed to work against HIV infection," observes Rick Young of the Whitehead Institute, "but it does leave us with some uncertainty."

There are, amid the gloom, a few hopeful signs and enough clues to stimulate new avenues of research. A decade into the epidemic, more than half a dozen research teams around the country have been working with many patients who were infected with HIV at roughly the same time. As Leslie Louie of the University of California at Berkeley reports, those patients haven't all reacted to infection the same way. "We know now," Louie tells us, "that people who are infected do not advance toward symptomatic stages of disease at the same rate, nor do they all develop the same symptoms. So there is *something* that is influencing the way the virus is causing disease in people."

Louie's data suggest that in AIDS, as in schistosomiasis, differences among genes that control the immune response may afford certain individuals greater resistance to HIV's attack. Other researchers have been puzzled by people whose sexual habits virtually ensure that they have been exposed to the virus, but who show no signs of being infected. Anthony Fauci, director of the National Institute of Allergy and Infectious Disease, was asked in 1992 by *Science* magazine whether he believed that there are individuals who have been exposed to HIV and remained uninfected. "There have to be," he replied. "There absolutely have to be. There's no way that's not the case." If that is indeed the case, studies of resistant individuals' immunological genetic programming could tell researchers which molecular "buttons" to push to create resistance in people not so genetically endowed.

There are other reasons to be hopeful about vaccines that may help infected people stay relatively healthy, even if they cannot protect them from being infected in the first place. "It is clear," Rick Young emphasizes, "that the immune system makes a substantial effort to destroy the virus and that it can recognize its protein components. In addition, we have learned that the loss of an active immune response in infected individuals can quickly lead to disease. Thus it appears that the immune system does hold the virus in check for at least some period of time."

For the present, molecular immunologists are concentrating on three main strategies: immunotherapy, therapeutic vaccines, and protective vaccines. All these approaches hold promise in two important findings: that the human immune system can make antibodies which prevent HIV from attaching to T-cell receptors, and that certain classes of T-cells seem to be powerful weapons in the battle—if they can be alerted and armed before being infected themselves. Researchers hope that both these defenses could be strengthened by a vaccine before infection or early on in the course of the disease. Two therapeutic vaccines have already shown some promise. At least some HIV-infected participants in ongoing clinical trials of HIV vaccines have managed to keep their T-cell counts near normal. Moreover, trials in chimpanzees of a vaccine aimed at the related simian immunodeficiency virus seem to have protected those animals from infection.

Still, because the natural course of HIV infection stretches over many years, evaluating vaccines is a slow, laborious process. "If it takes seven years for someone to get sick," Young pointed out in 1992, "then we'll just be starting to get our data in seven years. Ten years minimum if we start injecting people with a vaccine right now."

In the end, there is hope on many fronts: in the speed with which researchers are uncovering the immune system's molecular secrets, in the skill with which others are exposing pathogens' molecular strategies, and in the dedication with which clinicians are struggling to translate new information into treatments. But researchers know that they are still neophytes in understanding and manipulating the genes and molecules that enable our immune systems to operate. While looking to laboratories for cures and vaccines, we must remember that for years to come, the amount of human suffering and death inflicted by diseases ranging from schistosomiasis to AIDS will depend as much, or more, on the conscience, attitudes, and compassion of society as it will on molecular magic bullets.

6

The Sheep That Laid the Golden Egg

GENENTECH'S PRODUCTION FACILITY in south San Francisco resonates with a mechanical hum that seems to emanate from everywhere at once. Eight miles of gleaming stainless steel pipes snake through the huge main room, linking giant vats with a bewildering array of gauges, valves, wires, knobs, and computer screens. This enormous, expensive state-of-the-art device is called a bioreactor. Though many of its activities are regulated automatically, it is tended by a team of technicians in surgical suits who ensure that the fluid flowing through its metallic veins is maintained at the correct composition and temperature.

That precious fluid carries a stew of genetically altered cells and the products those cells have been engineered to produce—lifesaving human proteins that our bodies manufacture only in vanishingly small quantities. Depending on which cells the fluid contains and how they have been genetically reprogrammed, this bioreactor might hold the promise of a better life for diabetics, cancer patients, heart attack victims, or people who suffer from a wide range of genetic diseases.

Yet even as Genentech's production line shifts into high gear, the next generation of bioreactors is coming on line. Half a world away, in the Scottish town of Roslin, an employee of Pharmaceutical Proteins Limited (PPL) inspects a group of bleating, woolly, four-footed bioreactors that are sequestered in a high-tech barn to protect them from disease and injury. Although these sheep look no different from the common barnyard stock that sired them, they are among the most valuable animals in the world for, like genetically altered goats and cows being produced elsewhere, they carry pieces of human DNA grafted into their genome.

In creating these animals, PPL's genetic engineers have enlisted the complex, self-regulating cells, glands, and ducts of sheep mammary glands to replace the pipes and meters of artificial bioreactors. Under the direction of transplanted human genes, those mammary glands manufacture and secrete into milk the same sorts of human proteins produced by mechanical bioreactors, but with far less moment-by-moment supervision. Advocates for this approach assert that once engineered mammary gland production systems are fully understood and controlled, they will produce desired proteins more cheaply, more efficiently, and in greater quantities than the elaborate cell culture systems popular today.

Both these bioreactors herald a new era in our relationship with the living world. They testify that molecular biology now enables us not only to read the molecular texts that define and direct life, but to edit and rewrite them. Automated equipment and biochemical tools act like molecular word processors, "searching" for specific genetic "words" and "sentences," "cutting" them out of one piece of DNA, and "pasting" them onto another.

Although genetic engineering can produce astonishing results—often breathlessly described as designer plants and animals—this new development follows millennia during which humans have manipulated the genomes of other organisms by more commonplace methods. Selective breeding programs have created thousands of unusual plants and animals ranging from mainstream farm animals that produce more milk or thicker wool to bizarre varieties with characteristics far removed from their wild forebears.

Traditional plant and animal breeding, however, can work only *within* species, because barriers to natural gene transfer prevent, for example, mating a human with either a bacterium or a sheep.[1] Fantastic creatures made from parts of two or more species—horses with wings, or dragons with human heads—pepper the myths of many cultures precisely because those combinations defy the natural order of things. Such beasts are now generically called chimaeras, after the most famous of their ilk in Greek mythology.

Some modern chimaeric fantasies are no less misinformed than those of the ancient Greeks. "Carrot-Chomping Mom Bears Rabbit-Faced Boy," announced a tabloid headline not too long ago. The story, featuring an airbrushed photo of a toddler with floppy ears and whiskers, contained a quote from the distraught mother lamenting, "My doctor warned me not to eat so many carrots!" The implication that carrots have a direct connection with the characteristics of rabbits is nonsense to begin with. Further, the suggestion that a mother who eats too many carrots could somehow "push" the genes of her child in a rabbitty direction is ludicrously misguided. If rewriting the texts of life were anywhere near that simple, we would have been doing it long ago, and there would be no need for the expensive, painstaking, and time-consuming techniques of genetic engineering.

In reality, manipulating genes is both difficult and complicated. But increasingly sophisticated molecular tools are generating a steady stream of breakthroughs, enabling the transfer of human genes into cattle, insect genes into plants, and genes from almost anything into bacteria. The resulting *transgenic* organisms, though they may or may not look different to the naked eye, have attributes that even scientists imbue with an aura of mythic power; genetically engineered chromosomes which carry sequences from two or more species are described as *chimaeric DNA*.

The commercial applications of cutting and pasting genes spawned the biotechnology industry virtually overnight. In the process, it created a new partnership between basic research, which has developed essential tools through intensive and often dead-end experimentation, and business, which contributes its skill at developing and marketing products and maximizing production efficiency. The power of this merger is difficult to overstate, for medical applications have the potential to revolutionize health care, while agricultural biotechnology promises to produce a second "green revolution." Even a partial list of possible benefits from agricultural biotech is impressive: seeds with higher protein content, plants that require less fertilizer or none at all, crops that grow in salty soil, and plants that manufacture their own pesticides.

However, the revolution in genetic engineering has thrust itself upon us almost overnight, leaving little time to consider its implications. Genetic manipulation, offering possibilities limited only by skill and imagination, raises weighty scientific, social, and ethical questions. Thus far, partial information, misinformation, distrust, and fear have encumbered already difficult decisions on the uses of this new technology. Advocates and critics alike agree that this is the time for us to decide how we will use these new tools and how we will allow them to sculpt our fellow creatures, our planet, and ourselves.

"Biotechnology is a gift from the gods, but it's a gift we can use to destroy ourselves," warns Michael Fox, director of the Center for Respect of Life and Environment, an affiliate of the American Humane Society. "This is our last opportunity to use technology right. Hopefully, wisdom will help guide our species to use this new technology to heal the planet, to restore it, and to heal ourselves in the process. But if we're going to be indiscriminately putting genes all over the place, it's going to alter the way in which the biosphere, the ecosystem, and animals function. And there will be harmful consequences."

Jonathan MacQuitty of GenPharm, an international biotechnology firm, foresees an important role for biotechnology in our future: "I think biotechnology is going to significantly affect our relationship with the planet. I think it is going to lead to our having a much greater chance of influencing events, both to the good and to the bad. And I think it is going to bring this notion of responsibility to the forefront."

We cannot offer a thorough survey of biotechnology here; there are too many important developments to cover even a representative sample in passing. We will therefore concentrate on a few important stories relating to the transfer of genes from one organism to another. By looking first at selected medically related projects, then at others pertinent to agriculture, we hope to highlight the development of genetic engineering, the maturing science behind it, the opportunities it offers, and the problems it raises.

• • •

To ROBIN LOTT-LEDERMAN, Genentech's Protropin is a wonder drug. Since she began teaching her son, David, to inject the hormone into his bloodstream daily, she has seen marked improvement in his physical condition. Lott-Lederman is also delighted that the previously quiet boy has become more outspoken and confident. "I see in his face every day that he is happy about himself," she says with a smile. "It is worth everything to me."

David doesn't have cancer. He doesn't have a genetic deformity that causes passersby to notice him on the street. In virtually all respects, David is a perfectly healthy and normal teenager—to the extent that any teenager can be considered normal. The condition for which his family sought medical help is a subtle one. David is short for his age.

"He began to feel a little different because kids used to say 'Shorty' or call him names," his mother recalls. "And he just wanted to be like everybody else. They made him feel different, and I think that is one of the reasons why David was so quiet as a little boy. Some teachers treated him poorly because he was shorter. Because he was always quiet, he would be ignored in the classroom. People were telling David, 'Oh, you're so cute, but we won't pick you for this team.' The coach would automatically pick a kid one or two inches taller. So I just looked to the future and I saw David being a sixteen-year-old the size of an eight-year-old. He'd still be cute, but do you want to be cute at sixteen?"

Protropin, the drug that David takes in an effort to change this situation, is Genentech's brand name for a synthetic version of human growth hormone (hGH), which is normally produced by the pea-size pituitary gland at the base of the brain. Human growth hormone, along with other hormones, induces our bodies to grow during childhood. Some people, affected by clear-cut genetic defects or brain tumors, may not produce any hGH at all, or they may produce it in very small quantities. Such individuals remain abnormally short and may have other physiological complications that interfere with their health. But David doesn't lack growth hormone; there isn't anything medically wrong with him at all. It's not that his pituitary releases a pathologically low level of hGH; it's just that it releases somewhat less than an average amount. As a result, David, like many thousands of other children in his situation, would normally mature at shorter-than-average height.

David began lagging behind his peers in height early on, dropping off the bottom of the standard growth chart between the ages of four and seven. No one knew what, if anything, was wrong with him, and no physician could tell his mother whether or not David would continue to grow. From the information available, physicians estimate that the boy's adult height, if nothing were done, would likely have been between five foot two and five foot four.

That prediction bothered David and his family. "We wanted to give him the edge," his mother continues, citing studies dealing with societal perceptions of short people. "All things being equal, if there were two men going for a job interview, one of whom was six feet tall and the other five foot one, I think the six-footer would get the job. And I think that is how society sees people—what is on the outside of a person rather than what is on the inside. If there is a medication out there that can help kids like David, we want to give David that opportunity."

Robin Lott-Lederman can consider hGH supplementation for her child only because recent advances in genetic engineering guarantee an adequate supply of the hormone. The story of the urgent medical need for that supply and the advances enabling Genentech to produce it in their bioreactors raises important questions about the relations between biotechnology and the practice of medicine.

Until recently, there was barely enough hGH to treat even the relatively restricted number of children afflicted with severe growth hormone deficiency. Although the chemical composition of this precious substance was known, biochemists and chemical engineers could not devise a way to synthesize it. That problem wasn't and isn't unique to hGH, which is just one of a host of important, biologically active compounds that we can't manufacture with standard techniques of organic chemistry. The list includes hemoglobin, insulin, antibodies, interferons, and many more hormones and chemical messengers. We know what many of these molecules are made of, and we even know the base sequences of the genes that code for several of them, but we have yet to design an artificial process that accomplishes what the molecular machinery of the ribosome has been doing for millions of years.

Because of our inability to synthesize hGH, the entire medical supply was once obtained by extraction from the pituitary glands of human cadavers, a strategy that was problematical for two reasons. First, it requires hGH from roughly seventy cadavers to treat a single patient for a year. Second, cadavers were a potential source of infectious agents that attack the central nervous system. Because of that threat, the FDA forbade the sale of hGH from cadavers in 1985. Just a few months after the ban, however, Genentech made headlines by marketing Protropin, a version of hGH produced by genetically engineered bacteria.

The scientific foundations of this feat were established decades earlier. Scientists had been transferring genes from one bacterial strain into another and inducing those genes to produce their proteins since 1944. But that was just the beginning, because, as mentioned in Chapter 1, many genes from other organisms can also operate in bacterial cells. Although humans and bacteria have been separated by at least 3 billion years of evolution, our genetic codes are similar enough that their molecular machinery can still convert many, though not all, human genetic "sentences" into functional proteins. So it wasn't long before researchers began to try transplanting genes from other species into bacteria and getting them to work.

Today, selecting and isolating a piece of DNA and introducing it into bacteria is a simple procedure that even junior high school students can manage in a school laboratory with minimal equipment. The keys to its simplicity are molecular tools such as restriction enzymes and reverse transcriptase that molecular biologists co-opted from bacteria, and applied to dissecting and reassembling DNA. Over the years, researchers began isolating and identifying both the protein-coding parts of genes and the promoter regions that help control protein production. Once they could manipulate such segments of DNA dependably, researchers could take the coding region from one gene and attach it to the promoter region of another gene. That's how Genentech scientists engineered bacteria to make hGH: they spliced the human growth hormone gene onto a promoter that normally turns on a different gene in bacteria. They then inserted that whole piece of chimaeric DNA into bacterial cells.

But inducing the bacteria to assemble a human protein—a remarkable feat in itself—was just the first step in bringing a new drug to market. Bioreactors had to be designed and built to keep the drug-producing bacteria supplied with nutrients and to facilitate collection of the drug from culture vats. Then, before the drug could be approved and distributed, Genentech had to perfect techniques for isolating the hormone, purifying it, and proving that the product was safe.

Bill Young, vice president of production at Genentech, recalls participating in the first human trials of Protropin. "We did a clinical safety study with our first batch of Protropin. We thought it was 100 percent pure—the analytical methods available showed that to be the case. In those days we did initial safety studies with volunteers from the company, including me. We found out that patients who received it had reactions at the site of injection, and that led to an incredible amount of technology development which continues to this day. Protropin is now incredibly pure, and a lot of that [purity] in other products stems from what we learned in the early days as pioneers."

The result of their efforts was a safe and predictable source of supply that has made Protropin available both to seriously growth-hormone-deficient children and to normal but unusually short children like David. However, the productivity of bioreactors hasn't yet brought the price of Protropin down very far; the going price of a year's treatment for a child like David is close to $18,000. To gain maximum effect, many physicians recommend that treatment begin at least as early as age seven and continue until the onset of puberty.

"We know our products tend to be high priced," acknowledges Young, "but the technology is expensive, and the research is not cheap. A tremendous amount of technology goes into developing the processes that we use, and there are very expensive clinical trials. Many products do not ever make it to market, so the costs of research have to be recouped by eventual product sales." Eventually, Young and other industry representatives argue, the prices for Protropin and other genetically engineered drugs will come down, and the new products will be more widely affordable. In the meantime, certain health insurance policies cover the costs of such products if their use is medically indicated.

However, not all important human proteins can be produced by bacteria, despite the similarities between their genetic codes and ours. During the billions of years that our ancestors have been separated from the ancestors of bacteria, our cells have developed machinery that puts a range of sophisticated "finishing touches" on the proteins we produce. Mammalian cells often add sugars to amino acids after they have been strung together, for example, and frequently either "trim" protein chains or fold them in certain ways. Those essential modifications are not performed in bacterial cells. As a result, trying to produce some human proteins in bacteria is roughly equivalent to commissioning a bicycle factory to build high-performance cars; the tools in the shop just aren't up to the job.

One such complex protein is tissue plasminogen activator (t-PA), which can open the clogged arteries of heart attack victims. Properly administered, t-PA can often minimize, and sometimes completely prevent, damage to heart muscle following an attack. The market for t-PA is large, and companies like Genentech weren't willing to abandon it simply because their bacterial assembly lines couldn't handle it. If t-PA could only be produced by genetically engineered mammalian cells, researchers decided, they would learn to engineer and culture mammalian cells. "We looked around," Bill Young recalls, "and said, 'No one knows how to grow mammalian cells on any kind of scale, to make a product out of it.' That led to a kind of Manhattan Project effort that involved a third to half of the scientists in the company."

Their task wasn't easy. Mammalian cells grow more slowly than bacteria, are more delicate, and are much more demanding in their cultural requirements. Bacteria, after all, are independent, single-celled organisms, accustomed to fending for themselves. Mammalian cells, on the other hand, are pieces of larger organisms that depend on having their needs met by the body. What's more, the molecular machinery of mammalian cells varies from tissue to tissue, because fully differentiated cells in mature tissues are specialized to perform different tasks. Therefore, efforts to force a muscle cell to produce a protein normally made in the liver could be futile.

None of these difficulties proved to be insurmountable; the team at Genentech—and competing groups at other biotech companies—did develop mammalian cell culture systems. Soon those second-generation bioreactors were churning out t-PA, which Genentech brought to market in 1987. Another complex mammalian protein, erythropoietin (EPO), which regulates the production of red blood cells in the body, is being produced by similar methods, and a long list of other compounds is waiting in the wings—unless, that is, the third generation of bioreactors makes mammalian cell culture obsolete.

Those Scottish sheep we met earlier are at once a radically new phenomenon and the logical extension of other work in genetic engineering. Although mammalian cell bioreactors produce proteins that can't be made in bacteria, their yields are often low—typically about thirty thousandths of a gram per liter of fluid. Such quantities are sufficient to handle the need for potent proteins like growth hormone, but inadequate for others that must be used in larger quantities. As techniques of genetic engineering continued to improve, an obvious question presented itself: If some vital proteins can be produced only in mammalian cells, and if those cells are difficult to grow in artificial culture, why not induce mammalian genes and cells to do our bidding in their natural environments—the bodies of otherwise normal, intact animals? Animals are mobile, self-replicating, highly organized collections of mammalian cells that carry the added bonus of being protected from parasites by highly efficient immune systems. Why not engineer an entire organism so that its cells will produce the compounds of choice?

It was clear from the outset that not just any cell type would do. If genetically engineered cells produced a protein in an animal's blood, liver, or stomach, for example, either surgery or removal of blood or other body fluid would be a necessary—and cumbersome and debilitating—part of the harvesting procedure. On careful consideration, it became clear that mammary glands offer two major advantages. First, mammary gland cells normally produce proteins in abundance, so they seemed likely to carry the cellular and molecular machinery necessary to assemble a wide range of compounds. Second, mammary glands naturally deliver their proteins in a convenient form—milk—which can be easily collected with modified milking machines. Theoretically, all scientists had to do was insert the gene of interest into an animal that produced large quantities of milk and ensure that the gene was "turned on" in mammary gland cells.

Transforming that concept into reality, of course, entailed major practical difficulties, partly because farm animals are not the best experimental subjects for genetic engineering. Cows, sheep, and goats, because of their size, require more space than most established laboratories have available. Also, because these animals take so long to grow, mature, and bear offspring, each experiment takes several years.

Nonetheless, by 1991 independent teams in three different countries had succeeded with three different animals—cows, goats, and sheep, each of which is easily raised and produces substantial quantities of milk. Although none of those systems is yet in commercial production, the companies involved are happy with their progress. "The mammary gland is a very good factory," says Ron James, director of Pharmaceutical Proteins Limited, the Scottish company working with transgenic sheep. "Our sheep are little furry factories walking around in fields, and they do a superb job."

One product produced in PPL's ambulatory bioreactors is alpha-1-antitrypsin, a compound essential to the health and functioning of the lung. Patients with too little of this protein suffer from steady degradation of lung tissue, which leads to a form of emphysema. Replacement therapy by use of this drug has been approved in the United States, but supplies are inadequate. The compound is normally found in blood plasma, but extraction and purification from that natural source cannot meet the demand for it.

A ready market and restricted supply made this protein the perfect target for a genetically engineered production system. But because the molecule was too complex to be assembled by bacteria, either mammalian cell culture or a mammary gland–type system was clearly necessary. Deciding on the latter course, PPL set out to locate the gene for human alpha-1-antitrypsin, isolate it, and add it to the genome of an animal. That strategy required them to perfect two tightly interconnected procedures: the production of transgenic animals, and the creation of chimaeric DNA that, when used to transform those animals, performed as the researchers wanted it to. The story of their success, which illustrates both the increasing sophistication of genetic engineers and the bewildering complexity of the systems they work with, also reveals that their achievement owed nearly as much to artistry and intuition as it did to science.

To produce a transgenic animal, eggs and sperm are joined in a glass dish. Then, using tiny glass needles controlled by micromanipulators, researchers inject many copies of their chosen piece of DNA directly into the fertilized eggs before their first division into two cells. Eggs that survive the procedure, some of which have incorporated the gene into their chromosomes, are implanted into the wombs of surrogate mothers that have been prepared for unexpected pregnancy by hormone injections. Any of those eggs which have incorporated the foreign gene into their DNA will grow into animals which carry that gene in each and every one of their cells.[2]

Once this protocol for creating transgenic animals was refined enough to be dependable, researchers could concentrate on experiments that would lead them, step by step, to the production of chimaeric DNA that would do their bidding. Because farm animals are difficult experimental subjects on which to perform such experiments, most preliminary research was done with laboratory mice.

The first step—finding, isolating, and copying the human alpha-1-antitrypsin gene—was time-consuming but straightforward. Step two, on the other hand, was theoretically obvious but technically challenging. To make sheep produce alpha-1-antitrypsin in milk, the PPL team had to be certain that the human gene would be turned on, or expressed, in the mammary glands. That's tricky, because alpha-1-antitrypsin isn't normally produced by cells in that tissue. So the team had to attach their gene to a promoter sequence that is normally activated in mammary glands. In other words, they needed to find a genetic "phrase" that said essentially, "I will turn this gene on in mammary gland cells at the time milk is made."

Why was this a challenge? Researchers still don't know enough about genetic grammar to devise promoters from scratch. Therefore the PPL team couldn't simply create this type of DNA sequence; they had to find one which worked that way naturally. They managed to do so as follows. First, they isolated and identified a gene they knew was normally turned on in sheep mammary glands, the gene for beta-lactoglobulin. That protein is found in large quantities in sheep's milk, but not in mouse milk. One series of experiments transferred this sheep gene, with its promoter, into mice, and proved that it was indeed turned on in mouse mammary glands. (That was easy to determine, because this "non-mouse" protein appeared in the milk of their transgenic mice.)

Next, they took a piece of that sheep gene, including the region they hoped was the promoter, and successfully attached it to the human alpha-1-antitrypsin gene. They then inserted that artificial length of DNA into the mouse genome, hoping to "trick" the mouse mammary glands into treating the human gene like a milk protein gene.

Preliminary results were both positive and discouraging. When the team members analyzed milk from this new generation of transgenic mice they found that the animals' mammary glands were indeed producing alpha-1-antitrypsin. That was the good news. But the protein was being made at such low levels that it couldn't support a commercially viable production system. It wasn't at all clear why the system wasn't working as they had hoped it would. "We knew so little about how to put together a gene to make it work," recalls PPL researcher John Clark.

Because *some* alpha-1-antitrypsin was being produced, however, Clark and his colleagues knew that they weren't altogether on the wrong track. So they did what scientists do in such situations: they fiddled with their stitched-together gene, experimenting with different versions of it, and hoped that one or another version would produce more protein. Nothing seemed to work until one researcher tried an approach that had never before been used because it seemed counterintuitive.

Molecular biologists had known for years that mammalian genes are interrupted by puzzling sequences called *introns*—lengths of DNA that seemed to be little more than molecular gibberish separating parts of the genetic sentence. Because introns did not code for recognizable sequences of amino acids, scientists decided that they had no function at all, that they were just a type of "junk DNA." So when researchers assembled their artificial DNA sequences for genetic engineering purposes, they routinely created "streamlined" copies of genes, in which they edited out those bothersome introns.

Yet one of the strategies the PPL team decided to try in this case was to inject mouse eggs with a version of the alpha-1-antitrypsin gene that *included* some of its introns. The results were both surprising and gratifying. "We left some of these random bits of DNA in the gene, essentially as God provided it," says researcher Ron James, "and that produced high yields." In a way that is still not understood, the alpha-1-antitrypsin gene's introns somehow enhanced the production of protein.

PPL's John Clark remembers that great flush of success. "What can you do when you have made such a breakthrough? We went straight to the pub! I think the people there must have thought we were really crazy as we ordered yet another round to celebrate the achievement. Not so much *our* achievement, we felt, but the achievement of that particular mouse, who was a great mouse. We drank to the mouse, whose descendants are still going strong here."

That breakthrough mouse led to PPL's crop of transgenic sheep. Researchers injected 550 sheep eggs with the same sort of hybrid DNA sequences that had succeeded in mice. Four hundred thirty-nine of those eggs survived and were transplanted into ewe surrogate mothers, which ultimately gave birth to 112 lambs. Of those, only five, or less than 5 percent, had incorporated the human gene into their DNA. When these lambs matured, three of them began to produce alpha-1-antitrypsin in their milk. Two ewes deliver only three grams of protein per liter of milk, but Tracy, PPL's star animal, produces thirty grams per liter—the best rate of any transgenic farm animal in the world. PPL has already begun to use Tracy in breeding, and, as predicted, half of her offspring carry her winning genetic recipe for alpha-1-antitrypsin. This extraordinarily valuable ewe is truly a sheep that lays golden eggs.

But PPL's work isn't over. The company must learn to extract the human protein from sheep's milk efficiently and convince regulatory agencies that their product is pure and safe. Ron James admits that mechanical fermenters have an apparent edge over barnyard animals in this respect. "The fermenter's biggest advantage is that it is a very controlled system. It does the same thing again and again, and the regulatory people like that. We have to convince them that our sheep, which are not identical and won't always produce the same thing every day, are going to produce the protein consistently enough. Also that the purification process will produce it absolutely and consistently and identically."

Not surprisingly, PPL has rivals in the high-stakes world of farm animal genetic engineering. Its two biggest competitors have chosen to work with other animals. In Massachusetts, researchers in a joint effort of Genzyme and Tufts University hope to produce several proteins, including t-PA and a protein for treating cystic fibrosis patients, in goat's milk. In Holland, GenPharm is working to add a protein called lactoferrin, normally produced in human milk, to cow's milk.

Numerous other companies have found a wide range of medically relevant possibilities in transgenic livestock. In Princeton, New Jersey, for example, the DNX company is working on a human blood substitute produced in the blood cells of genetically engineered pigs. DNX researchers have already transferred the human hemoglobin gene into pigs and learned to isolate that human protein from pig blood. They plan to offer this crucial blood component as a way to circumvent concerns about

AIDS- and hepatitis-contaminated blood supplies and to minimize shortages caused by the limited shelf life of human blood.

As EXCITING AS developments in genetic engineering of animals are, they raise a number of ethical, legal, and social questions. Some have to do with the influence of the profit motive on research, others concern proper applications of new drugs whose uses have few precedents, and still others concern our use of animals for human benefit.

Consider the economic issues first. Genetic engineering requires intensive, cutting-edge research and development and involves several unproven technologies. As a result, this high-cost, high-risk business requires massive infusions of venture capital to get it going. Yet the prospect of enormous markets is enticing. By 1992, roughly eleven hundred biotechnological companies listed in the United States, many of them struggling start-up operations, accounted for $4 billion in annual sales. By the end of the decade, the market for biotechnology products is expected to reach $50 billion.

The promise of huge profits in an infant industry created a feeding frenzy among investors and drew participants from many university laboratories. But in this rush to commercialize genetic engineering, two closely related questions were largely overlooked: Is it appropriate for decades of research subsidized by public funding to produce profits for privately held corporations? More pointedly, is it wise to allow an industry spawned by public funds to be directed not by the needs of society, but by directives to maximize private profit?

Richard Godown, head of the Industrial Biotechnology Association, argues that private enterprise offers the most efficient way to commercialize and apply new technologies, and that profit is the only factor that encourages necessary investment and diligence. "Without the profit motive," Godown explains, "there wouldn't be a biotechnology industry, and the benefits the public is already enjoying from this new industry simply wouldn't be there. Biotechnology represents the best example of the free-enterprise system, in that entrepreneurs and scientists have gotten together to produce these wonderful discoveries. Everyone is benefiting. The economy is benefiting, and most especially, patients are benefiting."

Predictably, not everyone is so enthusiastic. "[Biotechnology] is profit-driven, not public interest–driven," warns Andrew Kimbrell, a litigator for the Foundation for Economic Trends. "It would be naive to say that everything developed for profit is against the public interest. But it would be equally naive to say that everything generated because of profit is good for the public interest." As an example, Kimbrell points to Genentech's sale of Protropin, which he feels is aggressively marketed to perfectly normal individuals. "[Genentech] has sold well over $150 million dollars of this drug in the last two years. You can't tell me that's just for dwarfs. These children are not ill, but are on the bottom three percent of height for their age. Parents have been convinced by drug companies, by doctors, to inject the steroid into their children even though there is a potential link to health impacts. The reason is a hard-sell job by pharmaceutical companies that are using this product on children. They had a cure—human growth hormone—but no disease. So they created shortness as a new disease, and are treating it."

The questions raised about Protropin are typical of many other issues related to genetic engineering. They are also disturbing, not simply because it is the nature of pharmaceutical companies to market products aggressively, but because they highlight an emerging new reality: molecular biology is giving us unprecedented power to alter the nature of human life. That new ability raises a host of ethical issues that have never before existed. Protropin, for example, is not sold over the counter. Doctors must write to Genentech certifying that their patients have clinically determined growth hormone deficiencies before the company will ship the drug. But David Lott-Lederman is not, as Kimbrell might say, a dwarf; he has only slightly lower levels of hGH in his bloodstream than a normal boy. He and his family are participating, with full informed consent, in an experimental study to determine whether Protropin can help people who are only slightly shorter than average to achieve a more "normal" height.

At present, Bill Young of Genentech insists, people like David can get the drug only if they are participating in such a study. Otherwise Protropin is given only to people with a medically defined condition. Young declares, "We send our people out to make sure that inventory control and security are appropriate, and that risk of theft is as low as we can make it. I think all those things go toward making sure that Protropin is used for the kids who need it. And the kids who need it *do* need it. Without Protropin they're going to be dwarfs. Very, very short, and not happy people."

The Human Growth Foundation, a national, nonprofit organization of health care professionals and parents of children with growth disorders, seems to agree with Robin Lott-Lederman's judgment that short stature can create difficulties for children. Its literature advises parents to "play an active role in monitoring their child's height progression through his or her growing years." While pointing out that extremely short stature can be an indicator of serious health problems, including kidney, heart, and lung disorders, the foundation's literature also emphasizes that short children can "develop serious emotional and psychological problems." Males, the foundation implies, may react to short stature "by taking on the role of a mascot with bigger children and a Mama's boy at home." Short females, the literature suggests, react with a tendency "to withdraw from others and use it as a way of maintaining childhood."

Opponents of this philosophy, who sometimes try to make an arbitrary distinction between "dwarfs" and short children, question the psychological repercussions for short children who are told they need medical attention to be "normal." Both arguments beg overriding questions with no simple answers: If new technology can alter the human condition in a way that apparently causes no harm and may have psychological benefits, on whom and under whose supervision should it be used? Does a decision to "treat" short stature with drugs or other therapies create, as Kimbrell suggests, a new "disease" from a state previously considered normal? If so, how will the public perceive people who do not choose or cannot afford such treatment? Finally, where does one draw the line?

Physicians affirm that the demand for Protropin has skyrocketed. "My colleagues and I were literally swamped with telephone calls," recalls Paul Saenger, David's physician. "Anxious parents wanted their boy or daughter to be not just average, but above average—so they could make a basketball team. That goes beyond medical indications for growth hormone and becomes merely cosmetic treatment. I think that is where physicians have to draw the line." Other doctors point out that as long as Protropin's cost remains high, physicians retain a certain degree of control over its use. But when the drug's price drops to a few hundred dollars a year, as it will eventually, demand will jump again when a wide range of people realize that they can afford Protropin treatment for their children.

In the end, this question dramatizes the relations between genes, our ability to manipulate genes, and our social environment. If society responds irrationally to a genetically related condition, and molecular biology makes it possible to change that condition, should we try to "cure" it? Or should we work to alter societal attitudes? Understanding the difficulty of changing society, and wishing to do what is best for their children, Robin Lott-Lederman and many other parents are making difficult decisions as best they can. "I think it's the world that is creating these problems," she reflects, "people making me think that there was something wrong with having a short child. It's not a perfect world, and people look at people who are different as odd—and handicapped, almost. I don't feel it should be like that, and kids aren't nice. But some adults aren't nice, either. As a mother, I didn't want my son to go through life with something hampering him."

Still another set of problems faces both academics and industrial scientists working with transgenic animals. In 1989, for example, animal rights activists in Scotland firebombed a laboratory. Luckily, no humans were killed, and the sheep, which were housed elsewhere, escaped harm. But PPL had its first taste of how strongly some people feel about tampering with animal genes. Animal activists—at least those who understand the realities of the research at places like PPL—don't usually get upset about the way transgenic farm animals are treated. That's not too surprising; an animal worth more than a million dollars is treated with the utmost care and respect by its owners, according to Jonathan MacQuitty, a spokesperson for GenPharm, a subsidiary of the company responsible for the first transgenic cattle in Europe. "If the animals wanted to eat caviar every day"—MacQuitty chuckles—"we could certainly justify the process and the cost involved."

What informed activists object to is the basic notion that humans can own and tamper with the genomes of other creatures. However, ownership of genetic material is definitely important to biotech companies. PPL, for example, wants to patent both the technology they developed for making transgenic sheep and the sheep themselves. Other companies working with transgenic goats and cows have applied for similar patent protection. The companies explain that patenting is the only means to protect corporate investment while facilitating the release of research information into the public domain so that others can build on their successes.

Critics like Andrew Kimbrell argue that living beings should not be patentable. "We have to be very clear," he insists, "that the animal kingdom, the plant kingdom, even our own bodies are not manmade products for sale. The market system would like to turn everything into a commodity. We are not commodities. There is a dignity, an integrity to life that should differentiate it from any commodity. We are looking at a world where animal species have been cloned and patented like machines, where our children will grow up looking at animals as no different from toasters or tennis rackets or other machines that we make."

"We have been using animals for millennia as part of farming," counters GenPharm's MacQuitty. "We have been doing it in a very unsophisticated process over the last thousand to two thousand years through animal breeding. What genetic engineering allows one to do is to make very precise changes, and to make very precise animals as a consequence."

"There is a small group of people for whom any use of animals appears to be wrong," admits PPL's Ron James. "The die-hard vegetarians say you shouldn't eat them, you shouldn't use wool that is collected from them. Anything at all. You can't argue with these people because there is no logic which will appeal to them." Genzyme's Alan Smith agrees that such people are hard to argue with. But in his view "transgenic animals are really just another extension of the uses of animals that people have found acceptable in the past."

In another vein, critics like Kimbrell argue that moving genes from one species to another blurs the identity and integrity of naturally occurring species. "The difficulty we have with creating transgenic animals is one of limits," Kimbrell explains. "We have put dozens of human genes into animals. The question is, *How many* genes can we put into animals? There is now only one to two percent difference between our genetic code and that of the primates. Is it okay to transfer some or all of that to primates? As we transfer human genes, perhaps sensitive human traits—intelligence, for instance—to the animal kingdom, what are the limits? And who's going to set those limits? Congress? Religious leaders? A popular referendum? We haven't even begun to think about those questions."

Biotechnology advocates insist that Kimbrell's group thinks up unrealistic scenarios to scare people away from the benefits of biotechnology. Says Jonathan MacQuitty, "The limits to this technology are more than one might imagine. The sorts of changes we are making to dairy cattle are a thousandth of one percent of the genome—very slight modifications that can bring us significant benefit. But a cow is a cow is a cow. We haven't changed things significantly from that point of view. I don't see wholesale changes being made for centuries."

• • •

COWS, SHEEP, GOATS, and other farm animals, however much attention they garner from the press, represent only one branch of agriculture and only one arm of agricultural biotechnology. In the global scheme of things, plants are farmers' primary emphasis, and nowhere is either the importance or the context of plant genetic engineering more obvious than in Holland. Mention of the Netherlands usually conjures images of windmills, girls in wooden shoes, boys on silver skates, and fields of tulips—in short, a sylvan way of life stalled comfortably in the middle of the last century. In reality, Holland is modern, densely populated, and intensely agricultural. It is also a nation that manages its present and faces its future in the context of a centuries-old need to alter and control its environment. The story of the boy whose finger prevented the collapse of a dike may be apocryphal, but it emphasizes the age-old Dutch struggle to wrest land from the North Sea.

Over the last several decades, dense population and limited space have led Dutch agriculturists to favor intensive management of agricultural production. Resulting yields are impressive. Holland is roughly the size of New Jersey, yet Dutch farmers maintain 4 million dairy cattle, nearly half the number raised in all fifty of the United States combined. Grain production is also outstanding. "In Holland we produce up to seven and a half to eight thousand kilograms of cereals per hectare," explains Dick Bÿloo of Florigene. "In the United States I believe it is about four thousand." Flowers, seeds, and bulbs are also serious business—the flower industry alone employs more than 70,000 workers, boasts 11,000 acres under glass, and funnels its produce through the largest wholesale cut-flower market in the world.

But the cost of sustaining this productivity is becoming apparent. To grow so much food on so little land, Holland has become addicted to agricultural chemicals. Its famous tulips, its productive grain fields, and its flower industry are hooked on chemical fertilizers and pesticides that taint soil and contaminate water. In certain parts of the Netherlands, plants are grown hydroponically in greenhouses, not by choice but out of necessity; the ground beneath them is contaminated with fertilizer salts and pesticide residues, and pests have evolved resistance to poisons that were once effective against them. In response to Holland's growing environmental dilemmas, the Dutch government has passed legislation requiring farmers to cut their use of pesticides, insecticides, and fungicides in half by the year 2000.

The combination of these serious practical imperatives and the Dutch fondness for technological innovation makes Holland the perfect environment for agricultural genetic engineering. As a result, this single small country sports examples that span the full range of goals for that technology, from apparently frivolous fiddling with the colors of cut flowers to obviously serious efforts to improve food production through engineered improvements in pest and disease resistance.

The most talked-about flowers in the industry in 1992 look innocent enough, and are unlikely to evoke exclamations of awe from recipients. From a distance, they seem to be little more than common white chrysanthemums. Yet these particular white mums are the products of genetic engineering, and are said to exhibit a purer, "whiter than white" display of petals than has ever been seen in their species. Created through a collaboration between the Dutch company Florigene and the American company DNAP, they were welcomed into the cut-flower trade with great fanfare. The patented plants that produce these blooms have been given a name that trumpets their developers' aspirations: Moneymaker.

Moneymaker was spawned by a business that faces competition as stiff as that in any other industry. International cut-flower markets are fickle and constantly shifting, so Dutch growers work diligently to keep up with changes in market preferences. "It's a problem," explains Florigene's Bÿloo. "The colors are very trendy. Even the breeders can't say, 'What colors do we want next year?' So if there is a possibility to switch very quickly—because we have the genes and can put them into flowers—we can go trendy with them to the market."

That's the reason for the excitement over Moneymaker and the much-heralded blue rose. No one suggests that these plants in and of themselves will revolutionize the flower industry; the feeling is rather that they signal possibilities to come. In the narrowest sense, growers look forward to being able to say "the Milanese design community wants yellow delphiniums next spring, and the Parisians want pink marigolds. The genes for those flower colors are in this DNA library. Get to it."

But in the broader sense, changes in flower color are only subtle hints of genetic engineering's potential to transform current agricultural practices. "Plant biotechnology is extremely important to the Netherlands," asserts Ben Dekker of Mogen, another Dutch plant genetics research firm, "since the Netherlands have a very important role in the seed industry. We think that biotechnology is the only way they can add value to their seed and that way sustain their position in the market."

Mogen is concentrating its efforts on developing crops resistant to pests that destroy 40 percent of the harvest every year, and even more during serious outbreaks. Fungi invade grapes; viruses sicken everything from roses to rice, and insects devour cotton plants. Until lately the only defense against these pests has been harsh chemicals, but because those are being proscribed, Dutch farmers—like many others around the world—must find new ways to protect their crops. To discover these alternatives, Mogen is trying to unlock the secrets of the plants themselves. "Plants do have natural defenses," explains Dekker. "They can't run away, so they have to defend themselves against fungi or insects or whatever. We have found a way in the lab to switch on this defense system."

Plants don't have the elaborate, nearly infinitely malleable system of antibodies and T-cells that many animals use to target specific parasites. Instead, plant defenses employ several less-understood systems that work more generally. Many plants synthesize compounds that work as natural insecticides—pyrethrum, extracted from a species of chrysanthemum, is one potent example. Other plants produce compounds that have similarly toxic or deterrent effects on other pests. Nonetheless, many crops, such as potatoes, are susceptible to fungal diseases whose devastating effects can have wide-ranging economic and social effects. The potato famine in Ireland, caused by unusually damp weather that allowed a fungus to wreak havoc on crops between 1845 and 1860, forced unprecedented numbers of peasant farmers to emigrate, many of them to the United States.

Mogen scientists have discovered a tobacco plant gene that confers resistance to certain fungal blights. "Now what we do," explains Dekker, "is modify the timing of the expression of those genes, taking care that the fungus is attacked before it can infect the plant." Mogen doesn't hold a patent for this resistance gene, but expects to profit from the proprietary technology used to transfer genes into specific crop plants.

The company is employing a different strategy against viral pathogens, which affect many crops in the developing world and which, until lately, no chemicals could combat. In this effort, the company identified and cloned a gene from the invading virus itself, a gene that codes for the protein coat. Somehow, by a process that is still not completely understood, inserting that gene into plants confers increased protection against further viral attacks.

You may wonder why biotechnology is so important in this context. Why can't growers just breed new strains of plants from old ones, as they have for hundreds of years? One answer requires only a single word: "speed." Genetic engineering allows companies to create new crops much more quickly than traditional breeding methods. Speed is of the essence in trying to bring a trait like pest resistance into agricultural crops, because many organisms that cause disease are constantly mutating in ways that allow them to "outrun" host defenses. "The normal process of breeding takes five to ten, sometimes twelve years," says Dick Bÿloo. "One of my friends has been spending his whole life in breeding potatoes resistant to [a particular] disease, and every time he had a potato which was resistant to that disease, he found that the disease had changed and he had to go on."

Another problem faced by breeders is that genes for disease resistance are often discovered in wild relatives of crop plants, which may be so far removed from their domestic cousins that it is difficult to breed them together. Furthermore, those resistance genes are part of a genome that often contains a host of traits considered undesirable in food plants, such as bitter taste, small seed or fruit size, or generally low yield. Ordinary tactics for breeding resistance into hybrid crops take so long because they usually require several generations of selective crossing to separate the resistance gene from those undesirable characteristics. Genetic engineering, on the other hand, theoretically makes it possible to pinpoint, excise, and transfer a single gene to an otherwise desirable variety.

In addition, transgenic technology enables researchers to employ useful genes from organisms totally unrelated to their crops. For example, scientists at several companies are working with a gene from the bacterium *Bacillus thuringiensis,* which produces a substance toxic to insects. Early efforts to transfer that gene into plants were successful, after a fashion; if the gene is expressed throughout the plant, the toxin is present in those parts destined for human consumption as well. But agricultural genetic engineers hope to create chimaeric plant DNA in much the same way that PPL's sheep-research team did; just as PPL's artificial gene is turned on only in mammary glands, the new version of the toxin-producing gene would be turned on only in leaves and other plant parts that are eaten by insects but not by humans. Proponents argue that these genetically engineered plants can reduce the use of harsh chemicals, with enormous potential benefits for the environment. "If we could introduce genes in order to make the plant resistant to pests," notes Bÿloo, "then we shouldn't spray anymore. So the environment is released from being contaminated."

DUTCH GENETIC ENGINEERS haven't restricted themselves to plants; they dream of engineering several other agricultural products for which Holland is famous, including milk and cheese. GenPharm's MacQuitty dreams of a future in which companies can genetically engineer cows to produce many different kinds of milk. "You are going to have a lot of different specialty milks—milks designed for infants, milks designed for older people, milks designed for hospital patients—all of which have a slightly different formulation." As proof of the company's progress toward this goal, MacQuitty proudly displays Hermann, a large, handsome, but otherwise ordinary-looking black and white bull who regards visitors with a look of mild interest. "He is the only animal that made it to the cover of an international scientific journal," MacQuitty tells us with pride.

Hermann carries the human gene for lactoferrin, a protein produced in human milk, but not in cow's milk. GenPharm wants to engineer cows that produce this protein, which they will then purify and sell in infant formulas. "There are a lot of women of childbearing age in the work force who, for a variety of reasons, find it very difficult to nurse infants," says MacQuitty. "What we would like to do is to provide some of the benefits of human milk in formula so that working mothers and infants can take advantage of them." The company also plans to use lactoferrin in foods for the elderly, because the protein seems to have some therapeutic properties.

"Hermann, the test-tube stud," as American newspaper headlines described him, received official permission to reproduce from the Dutch Parliament in 1992. More precisely, GenPharm obtained government permission to use Hermann's sperm in an artificial insemination program involving forty cows. In the future, GenPharm hopes similar programs combining genetic engineering and selective breeding will lead to a variety of genetically engineered milk products that won't have to be extracted and purified, because they will be consumed directly as food. "The glass of milk needs to be fiddled with," asserts MacQuitty. "This was not perfectly designed for human consumption. For example, people have been trying to remove fat from milk for some time. A better idea would be to start with milk which has less fat in it to begin with." In addition to low-fat milk "on draft," MacQuitty envisions a wide range of designer milks containing newly designed proteins that help make new kinds of cheese or increase the shelf life of milk itself.

At this writing, several genetically engineered food crops are on the way to market. In 1990, the Dutch company Gist-Brocades received approval from Britain to develop a genetically engineered yeast strain that causes bread to rise faster than normal strains do. This was the first time a food product involving live engineered organisms received approval for development anywhere in the world. In the United States, Calgene Corporation announced FLAVR SAVR™, a genetically engineered tomato. This variety, in which the action of an enzyme that causes softening and rotting has been blocked, remains firm in transport to market, even if it is not picked until vine-ripened. Normally, commercially raised tomatoes are picked while they are still green—and hard, and relatively tasteless—to protect them from damage in transport to market. The nonrotting trait, according to company spokespersons, should allow commercial growers to ship full-flavored, vine-ripened fruit that tastes less like the current crop of "red baseballs" and more like the home-grown varieties of yesteryear.

THE TECHNICAL HURDLES that must be overcome to produce genetically engineered foods on a commercial scale are formidable, yet they are only part of the effort required to bring those goods to market. For many companies, the biggest problem isn't a recalcitrant gene or an intractable ovum, but the formidable opposition of the antibiotechnology movement. Various members of that diverse, loosely coordinated, international assemblage of people oppose particular aspects of genetic engineering for a wide range of reasons.

At about the time Moneymaker chrysanthemums were welcomed by enthusiastic crowds in Holland, for example, a genetically engineered petunia created at the Max Planck Institute for Breeding Research in Germany was spurned and rejected by that country's environmental activists. Germany has the most restrictive laws on genetic engineering in all of Europe, a legacy of the Nazi years when a person's "genetic purity" could determine life or death. So wary have Germans been of this new technology that their courts prevented an insulin-producing plant from opening for lack of legislative controls on genetic engineering.

Elsewhere, in more legislatively permissive countries, critics have raised numerous objections to the speed with which genetically engineered products are heading to market. Animal welfare groups, as mentioned earlier, sometimes cite concerns—usually unfounded—about the welfare of transgenic animals. The Dutch Green Party and Holland's Society for the Protection of Animals, for example, used that argument in their unsuccessful bid to stop the use of Hermann's sperm in breeding programs. Other biotechnology "watchdog" groups object to nearly any use of the technology in agriculture on broader philosophical grounds.

The Foundation for Economic Trends (FET), for which Andrew Kimbrell litigates, is one such organization, at least according to its detractors. FET is the brainchild of articulate and outspoken Jeremy Rifkin, the biotechnology industry's most persistent gadfly. Rifkin has successfully derailed several major biotechnology projects, beginning with the famous "ice-minus" lawsuit against the University of California in 1983. FET's efforts in that case brought to a screeching halt a series of field trials planned to evaluate the efficacy of genetically engineered bacteria in inhibiting frost crystal formation on plants. Following several other successes, Rifkin and his foundation are now taking aim at FLAVR SAVR™ tomatoes and other transgenic foods. Their cleverly named Pure Food Campaign is aimed at forcing the Food and Drug Administration, which treats genetically engineered foods in much the same way as existing "whole foods," to issue more restrictive guidelines.

Rifkin insists that his efforts aren't meant to stop progress altogether, but rather to ensure that the public understands the social and ecological ramifications of new technologies before they are approved by scientists and legislators. Biotechnology representatives and policy analysts—and numerous well-informed journalists and academic scientists—don't agree. Many get apoplectic at the mere mention of Rifkin's name. One commentary in *Biotechnology* magazine, "Rifkin Resurgent," described him as "part pragmatist, part idealistic visionary, and part snapping turtle."[3] Advocates for biotechnology see Rifkin as a modern-day Luddite, a man whose demagoguery conceals a deep-rooted dislike of technology. According to Russ Hoyle, author of the *Biotechnology* article, Rifkin's ostensible concern for safety and other scientifically relevant issues is driven by a personal social and economic agenda: to oppose any technology that he feels threatens the survival of "friendly organic farming on traditional family farms."

Although many of Rifkin's opponents scorn him, few dismiss him. His skill at public relations makes it difficult for concerned but scientifically uninformed citizens to separate serious issues from rhetorical exaggerations and opposition to technological progress. In 1992, for example, Rifkin encouraged a group of prominent gourmet chefs to boycott genetically engineered foods. "I got in touch with Jeremy Rifkin because I wanted more information on genetic engineering," Rick Moonen, executive chef at New York's Water Club told *Biotechnology*. "I started speaking with more chefs and they asked for information. There was a strong opinion against genetic engineering in food, so Jeremy and I decided to hold a press conference. . . . I don't know all the issues Jeremy's working on, but I'm confident he's doing the right thing." Soon newspaper headlines were asking, "Can 'Frankenfoods' make it?"[4]

There is no question that genuine concerns about agricultural genetic engineering do exist; they would for any new technology. A number of eminently respectable critics of genetic engineering share certain legitimate concerns that Rifkin and his followers raise. Many environmental groups, for example, realize that agricultural genetic engineering has the potential, at least theoretically, to make agriculture more "environmentally friendly." These groups fear, however, that such uses will be few and far between in the absence of legislative controls, because, thus far, recombinant DNA technology has been under the de facto control of major agrochemical firms. The Sierra Club and several other advocacy groups agree that Calgene's rot-resistant tomato will succeed or fail based on consumer choice in the marketplace. But they argue that consumers won't have nearly as much ability to control success or failure of other corporate efforts, including what some see as attempts to "make the world safe for increased pesticide sales."

Calgene, for example, is working to insert genes for resistance to the herbicide Bromoxynil into cotton plants. Calgene's work is partially funded by the French chemical company Rhone-Poulenc, which just happens to distribute Bromoxynil under the trade name Buctril. Monsanto, meanwhile, is engineering resistance to its herbicide Round-Up into cotton, soybeans, and canola. Industry advocates insist that these efforts—and similar research under way at American Cyanamid, DuPont, Sandoz, Ciba-Geigy, and Hoechst, among others—will allow farmers to use more specifically targeted pesticides in the knowledge that crop plants won't be harmed. According to industry spokespersons, carefully targeted use will allow farmers to get by with smaller amounts of less toxic herbicides, thus sparing the environment.

But the National Wildlife Federation charges that planting Bromoxynil-proof cotton on half of America's cotton-growing acreage would double sales of that product. Other environmentalists point out that Bromoxynil is hardly risk-free; one product containing it was withdrawn after it was shown to be carcinogenic, and the compound is toxic to trout at a concentration of 50 parts per *billion*.[5] As for questions about dosages, *Chemical Week*, a trade publication, suggested that "farmers would then be willing to use even more of the weed killer, safe in the knowledge that their crops won't be damaged."

Hopefully, the chemical industry and environmentalists, aided by regulations of one sort or another—will find an acceptable compromise that combines Mogen-style engineering to increase plants' resistance to pests and disease with safer chemicals for use in emergencies. In the meantime, the most outspoken critics of agricultural genetic engineering fear that biotech will simply replace one form of pollution with another. "We're seeing biotechnology used to create new creatures never seen before on earth," Andrew Kimbrell argues. "For example, we are now genetically engineering fish to contain human genes, chicken genes, cow genes. We want them bigger, we want them to reproduce faster. Salmon are being engineered with human genes to grow bigger. If we put those fish in a pond—we release them into the environment—they will pollute the native gene pool. We are trying to protect the salmon across the world right now, but we would pollute that native gene pool forever.

"There is one thing about biological pollution," Kimbrell continues. "You cannot recall it. You can never get it back once it's out there. We have an age in which we are concerned with chemical pollution. Twenty or thirty years down the line, we will have irretrievable biological pollution as we put these new genetically engineered creatures into the wild and destroy genetic diversity around the world, perhaps forever."

That problem would seem to be particularly pronounced with genetically engineered plants. "The greatest danger posed by genetic engineering," suggested an article in the *Arizona Republic*, "may come from sex with weeds." That may sound amusing but it is, in fact, a serious worry. After all, couldn't wind or a wandering bee carry pollen containing chimaeric DNA to related plants in the wild? Once the new gene gets into the wild gene pool, what happens? Could pest and herbicide resistance be carried to agriculturally destructive weeds? Could plants with the new gene outgrow their normal cousins? Could something which seems as harmless as a gene for flower color cause lasting damage to the ecosystem?

So far, the biotechnology industry has been able to address this type of technical problem, although not always for reasons that critics might like. For example, if "designer genes" could migrate out of engineered crop plants, seed companies that spent millions to develop those genes would be unable to protect their investment. One system under development to prevent the "escape" of patented genes on wandering pollen involves chimaeric DNA stitched together from several organisms and hooked onto a promoter region that operates only in developing pollen grains. When that system is turned on, it neatly "snips" the engineered gene right out of the plant's DNA and "stitches" the chromosome back together again. Thus, any pollen that leaves the plant will not carry the protected gene.

In addition, plant genetic engineers argue that their trade can help ensure genetic diversity in crop plants, at the same time making it worthwhile for major agricultural firms to preserve wild relatives of domestic species. "It allows you to keep the genetic base of crops as wide as possible," explains Robert Goldberg of the University of California at Los Angeles, "because you are not faced with a breeding situation where you have to keep backcrossing, backcrossing, backcrossing, to maintain desirable traits. What you can do is take your variety A, and put into it the seven genes you want it to have. You take variety Z, put the same seven genes into it. You can do that with one hundred varieties. Now you have kept the natural genetic diversity of the population, and, in fact, you have enhanced it. That's why I see this as being so positive, because instead of narrowing genetic bases by breeding, we are enhancing the genetic bases by keeping everything that is there naturally and then adding some stuff that makes it a little bit better."

Many environmentalists know that genetic engineering has potential benefits, as well as costs, if applied properly. "We are living in an increasingly dysfunctional planet today," notes the Center for Respect of Life and Environment's Michael Fox. "The aberrations in the climate, the ozone hole, global warming, acid rain, all of these things. The way in which we live is out of balance. How can we bring that balance back? How can we even think in a right-minded way to get biotechnology on the right track? These are the kinds of fundamental questions, the ethical and moral questions, that the biotechnology industry and policymakers need to address, rather than short-sighted views of profit and the illusion of progress. If we're going to use biotechnology right—and I think this is our last opportunity to do that—we have to have the right state of mind, the right attitude toward the rest of creation. I think that attitude entails a high degree of respect for how ecosystems work and how other life forms function. We need a greater respect and reverence for all life and the life processes of this planet that sustain us all."

One reason that plant genetic engineers are ahead of their colleagues in animal research is that plants, by virtue of their biology, offer a major advantage to gene jockeys. Many types of plants grow from relatively undifferentiated cells that researchers can culture in special nutrient media. As a result, it is often possible, by carefully manipulating nutrients and plant hormones, to culture a clump of plant cells, cut it into pieces, and grow entire new plants from each of those pieces. This process, both scientifically and popularly known as cloning, enables plant biologists to produce a nearly infinite number of genetically identical plants from just a few cells.

The advantages of this plant characteristic in the production of transgenic organisms is straightforward. While researchers can do their best to transform a culture of plant cells with chimaeric DNA, they usually succeed with only a few of them. But if they can identify those few cells, they can "clone" them to produce as many plants as they need. Animal genetic engineers have always been envious of that ability. If GenPharm could clone Hermann, or if PPL could clone Tracy, its star transgenic sheep, their work would be a lot easier. But animal cells don't work that way. It's true that even adult animals contain several kinds of stem cells that can mature and differentiate into a variety of different cell types. None of them, however, can be grown in culture to produce an entirely new animal—at least not yet. But two developments linked to a single laboratory promise to revolutionize genetic engineering in animals by providing a tool almost that powerful.

The man behind that tool, Mario Capecchi, doesn't look like someone about to change the world. Soft-spoken and thoughtful, he leads a quiet life, shuttling between his laboratory at the University of Utah and his secluded house overlooking snow-covered mountains. He needs the quiet and beauty of this home, he explains, to clear his mind, to offer him a different perspective on work in the lab. Often this process is so successful that he awakens his wife at two o'clock in the morning to discuss a new inspiration or idea.

Capecchi has broken through some of the long-standing barriers involved in animal genetic engineering. His discoveries make it possible to alter the genome of an extremely complicated animal, the mouse, with a precision heretofore unknown. Instead of simply adding a gene to a mouse and having it integrate randomly in the chromosomes, he devised a new strategy that enables him to program injected DNA to replace a specifically selected stretch of mouse DNA. What works in the mouse is liable to work in humans sooner or later. "The mouse genome is an enormous text," Capecchi explains, "consisting of a thousand volumes, each a thousand pages long. What this technique allows you to do is specifically go to volume 33, page 620, pick out a sentence, and change it at will. So we have essentially worked like a word processor."

Geneticists have been able to replace genes in yeast cells for years without having to work too hard. When a gene that resembled a normal yeast gene was placed into the cell, the added DNA would often switch places with its counterpart on the chromosomes. This process, *homologous recombination,* is so called because similar, or homologous, pieces of DNA interact with one another. This advantage, which made yeast researchers the envy of peers working with other organisms, enabled geneticists to "knock out" any single gene from a yeast cell and observe the effect of its removal. Unfortunately, no other organism seemed to possess the cellular machinery to swap pieces of DNA so efficiently. If DNA added to a worm or fly cell integrated into chromosomes at all, it seemed to insert itself randomly at locations possessing completely unrelated DNA sequences.

Capecchi's breakthrough was the discovery that mouse cells could also perform homologous recombination, albeit to a much lesser degree than yeast. Stunned when he stumbled on this ability, he shrewdly realized that he could take advantage of it to create the mouse of his dreams. "The cell has that capability to replace DNA on its own, and we don't have to invent it. All we have to know is that the capacity is there, and then exploit it."

The problem was to figure out how to get the gene of interest into a mouse cell, then grow that cell to produce an entire mouse. If that could be done, every cell of mouse produced by that egg would have the desired genetic alteration. Capecchi had already learned to use retroviruses to carry DNA into mouse eggs, essentially co-opting the viruses' genetic instructions for entering and infecting cells. To use that molecular intelligence safely, Capecchi "crippled" the viruses' normal activity by removing the particular genes that enabled the virus to multiply in the cell. He then inserted the mouse gene of interest into the remaining retroviral genome and allowed the tamed, engineered virus to infect mammalian cells as it normally does.

Capecchi found that genes so transferred into mouse eggs incorporated with high frequency into the chromosomes. But only a small percentage of these eggs—about one in 1,000 or one in 10,000—had actually replaced the corresponding gene in the egg DNA. Trying to implant enough of those eggs into surrogate mothers and grow them all to maturity would take far too long and use far too many mice to be practical. For that reason, he needed to find mouse cells with a special combination of features that egg cells lack. His ideal cells should be able to form an entire mouse as egg cells can. But the cells Capecchi was looking for, unlike egg cells, could survive in a culture dish long enough for him to find the very small number of them in which the correct gene replacement had occurred.

Martin Evans of Cambridge University in England solved that problem. Evans managed to isolate embryonal stem cells—ES cells for short—from very young mouse embryos. ES cells can be cloned and grown in culture, although they cannot be stimulated to produce entire new mice as certain plant cells can. But if ES cells are injected into a mouse embryo of the right age, they spread throughout the embryo and participate in the formation of virtually all the tissues of the resulting animal, including the germ line cells that produce eggs and sperm.

This method has opened up new vistas in mouse developmental biology; some boosters say that Capecchi has single-handedly made mouse genetics possible. Because mice take so long to grow and reproduce, it is difficult to find mouse mutants through the methods of classical genetics. Capecchi's method enables scientists simply to create their own mutants with genes originally cloned in other animals, including humans. Capecchi has already used this technique to observe what happens when a mouse lacks one particular oncogene from conception and when it lacks a particular homeobox gene.

Equally important is that this technique opens up a host of options to genetic engineers. One application is creating mouse "models" of human diseases. By eliminating the mouse equivalent of a gene known to be involved in a human genetic disease, scientists can create mice that can be used to test treatments for that disease. Paul Schmitt of DNX, a company that sells such models, explains his product. "Mice don't get Alzheimer's, they don't get high cholesterol. But we can genetically engineer these conditions into an animal so you have a whole animal system other than a human being on which to measure and evaluate drugs. This tool, before we had genetically engineered animals, was unavailable. It will accelerate the drug development process. Think how valuable that tool can be to any drug-testing or drug-development company."

GenPharm has already applied for patents on one transgenic mouse it has produced, a strain that representatives call MagicMouse. In those animals, its research team replaced nearly all the genes that produce mouse antibodies with the genes that produce human antibodies. The section of DNA that was swapped is millions of base pairs long, a feat of genetic engineering unmatched in mouse history. The result is a mouse that produces *human* antibodies when challenged with foreign proteins. GenPharm hopes to use MagicMouse to produce a variety of different human antibodies for use in medical treatments, including cancer therapies such as those discussed in Chapter 4, with an ease never before possible. If this strain is ever truly perfected—if it is shown to contain *all* the components of the human immune system—it will rival Mike McCune's SCID-hu mouse as a model for testing AIDS drugs and vaccines.

"We see this antibody-producing mouse being helpful in a number of different diseases," says Jonathan MacQuitty of GenPharm. "Bacterial infections, viral infections, AIDS, cancer." "This is the world's most valuable mouse," agrees MacQuitty's colleague Andy Barnes. "It has the ability to make any human antibody—it's your little factory making human drugs. And it is not making just a specific human drug for the rest of its life, like the cows and sheep. This is a way to instantaneously generate thousands of new human drugs."

Thus far, Capecchi's technique cannot be used in farm animals, although scientists are working on it. But that hasn't stopped companies from dreaming about what the future will hold. "Say you wanted, for example, to delete a protein in cow's milk," muses MacQuitty. "Suppose there was a protein that was allergenic—for some reason humans didn't tolerate it very well. You could go into the bovine genome and precisely remove the gene coding for that protein. Then you could make a milk which was much more palatable for human consumption."

Physicians at Massachusetts General Hospital in Boston have their own wish for this new technology. They are currently breeding miniature pigs for use as organ donors for humans. The group hopes to eliminate the genes responsible for the rejection of organ transplants, thereby enabling doctors to transfer pig livers, hearts, kidneys—any organ at all—into patients whose lives depend on new parts.

Of course, behind the euphoria over MagicMouse and its anticipated successors is a realization that is at once exciting and frightening. It fuels the wildest hopes of physicians involved with genetic diseases while it stirs profound anxieties among those who fear the ghosts of eugenics past. For techniques developed to create transgenic animals can ultimately be adapted for use on all animals, including humans. The promise and the perils of those applications are the subjects of the next chapter.

7

Redesigning Our Children

When Tessie Ashe was eighteen years old, she watched her nine-year-old niece, Farrell, die of cystic fibrosis (CF). In the grip of that fatal genetic disorder, Farrell spent much of her brief life fighting for every breath she drew. Her physical problem was as simple as it was implacable: she was being strangled by her own lungs. The cells lining her respiratory tract, their normal function impaired by a defective gene, secreted a thick, viscous mucus that blocked the flow of air. Farrell was so dependent on supplemental oxygen that she gasped and struggled like a fish out of water during the few seconds it took her mother to switch her mask to a new supply tank. By the time she succumbed to the disease, she weighed barely forty-nine pounds.

As traumatic as Farrell's death was, her suffering upset her aunt Tessie Ashe even more. Being forced to watch helplessly while the child lived in constant pain was almost more than Ashe could bear. So years later, when their son, Paul, was born, Tessie and Kevin Ashe banished from their minds the possibility that he could have the disease. Their optimism seemed reasonable; their first child, Emily, was a perfectly healthy two-year-old.

But Paul failed to gain weight normally despite constant feeding and had serious digestive problems and salty-tasting skin, all hallmarks of cystic fibrosis. Within weeks, the anxious parents found themselves in the all too familiar environs of the hospital where Farrell had died. When they received the awful news that Paul, too, had CF, Tessie Ashe felt as though she had been punched in the stomach. "You've just given my baby a death sentence!" she cried at the pediatrician. Sadly, the Ashes have a great deal of company in their distress; cystic fibrosis is one of the most common inherited diseases, striking one in every twenty-five hundred Caucasian newborns in the United States.

Symptoms of CF, like those of sickle-cell anemia, appear only in individuals burdened with two copies of a mutant gene.[1] People like Tessie and Kevin Ashe, who carry one normal and one defective copy, may never learn of the danger that lurks in their DNA unless they marry one another and have children. In those situations, half the mother's eggs and half the father's sperm carry the deadly allele. The probability that an unlucky egg will meet a similarly unlucky sperm is 25 percent, so each child conceived by CF carriers has a one in four chance of getting cystic fibrosis.

Thus Paul's diagnosis was a diagnosis for the entire Ashe family; it meant that both Tessie and Kevin were CF carriers and that any future child of theirs would face those odds in a potentially lethal round of genetic Russian roulette. That made the blow even harder to take, for the Ashes had dreamed of a large family. But under the circumstances, they would never try to conceive again. "The natural course is for my children to bury me," says Kevin Ashe. "I'd like to keep it that way."

Then, two weeks after the Ashes received the grim news about Paul, a scientific breakthrough lifted the pall of gloom that had settled over their future. In what the Ashes see as the work of God, researchers located and identified the gene that causes cystic fibrosis. Suddenly their family was thrust into the high-technology world of molecular genetics—a world in which constant advances raise both hopes about medical miracles and fears of social abuses. Some of those advances are already enabling the Ashes to fulfill their dream of a large, healthy family. Moreover, several lines of research, though still in experimental stages, hold the promise of effective treatment and possibly even a cure for CF during Paul's lifetime.

Paul's future, like that of his unborn brothers and sisters, is being rewritten as researchers learn to recognize and correct errors in the genetic code. As we described in Chapter 6, genetic engineers are beginning to read and edit genes, to assemble artificial stretches of DNA, and to use those "designer genes" to modify the body processes of bacteria, plants, and animals, including our fellow mammals. But genetic operations that work in mice can also, from a technical standpoint at least, be performed on humans. Beginning in 1990, several related technological breakthroughs, together with enabling regulatory decisions, have ushered in the era of human genetic engineering, also called gene therapy.

Even normally reticent clinical researchers herald human gene therapy as the first hope of true cures for previously intractable inherited diseases. Manipulations of the human genome have an astonishingly broad range of potential applications, from correcting genetic defects in the lungs of children like Paul, to removing that defect from the egg and sperm cells of Tessie and Kevin Ashe's future children, to tinkering with an embryo's genome to "enhance" normal characteristics, to engineering wholesale changes in the gene pool of the human race.

Given that intimidating array of possibilities, it isn't surprising that the emerging field is championed by overzealous proponents and plagued by equally ardent critics. Some advocates trumpet unrealistic utopian visions, implying that gene therapy can eliminate a great deal of pain and suffering from the human condition with few negative side effects. Meanwhile, opponents with apocalyptic fears raise specters of genetic discrimination and threats to the essence of being human. Both extremes pose dangers: the former encourages naive complacency that could foster abuse of technology, while the latter could seriously hinder the progress of medicine.

The ethical, legal, and social questions around human genetic engineering and its attendant technology are legion. How should testing for genetic disease be controlled? Who should have access to the information it provides? To what uses can that information be put? Is there any way to distinguish between gene therapy aimed at fatal disease and "enhancement engineering"—the genetic equivalent of cosmetic surgery? Who should set guidelines for the development and application of this technology? Should any limits be imposed at all? None of these questions have simple answers, as the Ashes have discovered.

• • •

PAUL ASHE'S LIFE has been surprisingly normal so far, because as yet he has only a mild form of cystic fibrosis. Living with his family in a sprawling brick house outside of Houston, he receives the standard treatment for people with his condition: enzyme supplements at every meal and a treatment called percussive therapy. Once a day, to loosen the mucus in Paul's lungs and help him breathe, one of his parents places Paul across their knees, and pounds his back and chest with cupped hands. To make the therapy as enjoyable as possible, they often strike him in rhythm, so that the clapping sounds match the pacing of Bible songs and gospel hymns they sing together. To make the treatment a family affair, the Ashes encourage four-and-a-half-year-old Emily to perform a similar treatment on her dolls.

Despite their loving attention and top-notch medical care, Tessie and Kevin are afraid that the mercurial disease could take a turn for the worse at any moment. "Right now he looks great," Tessie Ashe observes. "But with cystic fibrosis it's progressive. It could get where his lungs are involved, it could get where he has liver involvement. We would have to worry about heart/lung transplants. All that is facing us."

As recently as a decade ago, Paul's prospects would have been grim indeed. In the past, CF patients rarely reached their thirtieth birthday, although new treatments have helped prolong their life expectancy. But today there is hope, thanks to Francis Collins of the University of Michigan and Lap Chee Tsui and Jack Riordan of the University of Toronto, who in 1989 discovered the gene for CF. "Making predictions about the timing of a cure puts one on thin ice because of the uncertainty of so many steps," says Collins, dutifully qualifying his enthusiasm. "But we hope that in the lifetime of kids who are alive today with CF, we will come up with something that will be beneficial to them—perhaps even a cure."

Few people understand the magnitude of recent strides in the fight against CF as well as Francis Collins, and few understand so well the problems and frustrations ahead for both researchers and patients. Collins, a consummate gene hunter, uncovered two human disease genes by his fortieth birthday, a research feat equivalent to winning two gold medals in two successive Olympics. He admits that the search for the CF gene was especially difficult; it took roughly eight years, was fraught with disappointments, and cost an estimated $50 million.

Why was finding this particular gene so difficult? "Trying to find the cystic fibrosis gene," Collins explains, "is like trying to find a single burned-out light bulb in the basement of a house somewhere in some town in the United States. Your original problem is that you don't even know which state to look in. Finding the right chromosome is like finding the right state. Narrowing the precise interval on that chromosome where the gene should be is like sifting through the counties and towns. Then you start a house-to-house search."

That search requires scientists to examine each and every piece of DNA in the suspected area of the chromosome which seems to code for a protein. During the examination, gene hunters search for any protein-coding region that seems to be different in CF patients but not in non-CF patients; those regions are likely candidates for the burned-out light bulb. "It is a roller-coaster experience," Collins recalls. "One day you are up, you think you have a great clue, you are just about to figure it out. The next day you realize that clue was all wrong and you are down in the depths again. You look in a *lot* of houses, finding that there is nothing wrong with their light bulbs."

The CF mutation was particularly elusive because, as it turned out, the most common error in the gene involved changes in only three base pairs. "Some other genes that have been found," Collins observes, "like muscular dystrophy, or fragile X syndrome, were large genetic rearrangements that you could see from a mile away, like a neon sign in one of the towns telling you, 'Look at this house!' CF did not give us that luxury."

Another problem was the lack of a map for the region of the chromosome harboring the CF gene. As a result, Collins and his colleagues had to search for their defective bulb without a list of house numbers, a map of local streets, or even a county map to show where the towns were. They had to build their maps as they went along. Advocates for the Human Genome Project note that their ongoing effort to map and sequence all our genes will make future gene hunts such as this one a great deal easier, and far less expensive. Lee Hood, for example, suggests that if the Human Genome Project had been completed before the search for the CF gene began, the gene's $50 million price tag could have been cut to somewhere around $200,000.

Why is finding the gene so important? First of all, finding and "reading" the *mutant* form of such a gene enables researchers to identify differences between the gene's normal protein product and the defective version that causes disease. The product of the normal form of the CF gene, for example, is a protein called CFTR, which acts to control the flow of chloride ions across the cell membrane. That pump is destroyed in CF patients, causing, among other things, their salty-tasting skin and the buildup of mucus in the lungs. Recently, researchers have identified the specific changes in amino acid sequence in the mutant protein, a useful step toward developing an effective drug treatment. Collins has identified a drug that can activate CFTR in cystic fibrosis patients, and hopes to initiate human trials.[2]

Second, identifying the *normal* form of a gene enables experiments in gene therapy aimed at curing the condition by replacing the faulty copy. Already, two separate research teams have been able to "cure" individual cells of CF in culture dishes by providing them with normal copies of the gene. One team has gone a step further by successfully introducing normal copies of the CF gene into the lungs of rats and by demonstrating that transformed cells in the animals' lungs produce CFTR. Efforts are under way to apply that technique to CF patients. We will describe them later in the chapter.

Third, knowing the precise sequence of the mutant allele enables the development of tests and other practices that may prevent the disease from being passed to future generations. This "transgenerational" intervention can take several forms, ranging from genetic screening programs that identify and offer reproductive counseling to therapy aimed at eliminating such alleles from the so-called germ-line cells that produce eggs and sperm.

The most clear-cut medical technique for avoiding the birth of children with genetic diseases—and the only one available until recently—is both physically and psychologically traumatic. The procedure involves collecting fetal cells from the womb by amniocentesis eighteen weeks into pregnancy, and performing a genetic test for the mutant gene on those cells. Should the test reveal that the embryo carries the deadly mutation, abortion is considered. That course of action, however, was never open to the Ashes, whose religious beliefs include unshakable adherence to an anti-abortion philosophy. "We would not abort," Tessie Ashe asserts emphatically. "That's not an option for us."

But new, extraordinarily sensitive genetic testing methods, facilitated by the discovery of the CF gene, offer Tessie and Kevin the hope of bearing children free of cystic fibrosis through a procedure with which they can personally feel more comfortable. The technique, *preimplantation diagnosis,* is still experimental and not predictable enough to guarantee success. But the Ashes have traveled to London to join the handful of willing experimental subjects at Hammersmith Hospital. "To have a healthy child," Tessie affirms, "I would go to the ends of the earth."

The key to the Hammersmith procedure is the hospital's highly respected *in vitro fertilization* (IVF) clinic, originally set up to serve infertile couples. Normally, the female member undergoes several weeks of hormone treatments, which induce her ovaries to release numerous eggs. Those eggs are harvested, then joined with the husband's sperm in a glass dish. Viable embryos selected from the resulting fertilized eggs are then implanted into the mother's uterus.

The Ashes are hardly infertile—"I think about becoming pregnant and I become pregnant," laughs Tessie. But they, too, will undergo in vitro fertilization, with the addition of one critical step. Their fertilized eggs, rather than being implanted at random, will be incubated for three days, coddled in culture as they grow into clusters of eight cells. Those eight-cell embryos then become the subjects of what is arguably the most delicate operation ever performed on human tissues. First, microscopic instruments are used to bore a tiny hole in the protective layer surrounding the embryo. Then an equally microscopic syringe removes a single cell, which is rushed to the laboratory for genetic testing. Using the PCR technique, lab personnel make millions of copies of that cell's DNA, providing enough genetic material to facilitate a search for the CF mutation in the embryo's genome. Only embryos determined to be free of CF will be implanted into Tessie's uterus. There, with luck, at least one will develop into a normal baby.

Intriguingly, the PCR technique that makes this part of preimplantation diagnosis possible is also the procedure's potential Achilles' heel. PCR's amplification of DNA is so sensitive that if even a single cell were to fall from a technician's fingers into the test tube, foreign DNA could contaminate the multiplied DNA and invalidate the results. For this reason, blood cells collected from the Ashes and several relatives were used to categorize the family's genetic makeup. DNA fingerprints made from those cell samples were compared with DNA taken from the embryos in an effort to ascertain whether or not such contamination had occurred. These precautions are vital: in one earlier effort, using less stringent procedures, the Hammersmith program improperly diagnosed the gender of an embryo sired by parents carrying a sex-linked disorder, and the mother bore a child with a serious genetic disease.

The current version of the procedure was developed by researchers Alan Handyside, Robert Winston, and several other colleagues. Winston, director of the IVF program, is a solemn man with a lined forehead and frizzy black hair that stands out in tufts on his head. In an office covered with photographs of babies born through the IVF program, he describes how years of animal experiments convinced him that removing a single cell from an eight-celled embryo is safe.

"To prove that it is harmless, the procedure is being done in a number of different animals, and there have been no embryonic defects. All the evidence we have is that each cell in any mammalian embryo, including the human, is totipotential. That is to say, if you divide an eight-cell embryo into eight cells, you can have eight identical children, each with a complete set of limbs, organs, brain, and everything. Obviously, that would be totally unethical and nobody would dream of doing it. But by removing one cell—if you like, by sacrificing one cell—you don't alter the status of the embryo."

By 1992, when Tessie and Kevin Ashe arrived at the clinic, six perfectly normal babies had been born after undergoing preimplantation diagnosis for other genetic diseases. Buoyed by that information, and after a great deal of thought and heartfelt prayer, the Ashes decided that preimplantation diagnosis was not only their best option, but that it was also consistent with their religious convictions.

For one thing, the Ashes have been assured that up to 80 percent of human embryos at the eight-cell stage don't implant successfully in the uterine wall, even under completely natural conditions. Those embryos are normally sloughed off by the body without their would-be mother even knowing that any of her eggs had been fertilized. In a sense, therefore, the selection process of preimplantation diagnosis occurs at, or before, the stage of development during which a similar, though more random, weeding-out process occurs in nature.

Furthermore, because each of the eight cells in this early embryonic cluster is totally undifferentiated, the Ashes can think about them very differently than they would about the parts of an embryo further along in development. "I feel comfortable with this," Tessie explains, "because each cell does not know where it is going to be. It's not as if there is a beating heart or a brain wave, so at this point I feel more comfortable with it than I would terminating a pregnancy."

Numerous other members of the large and diverse prolife movement, however, disagree with this rational assessment. Such individuals believe that human life is sacred from the moment of conception onward and insist that no interference with the reproductive process is permissible. What such people would say to parents of children suffering with CF is unclear. Whether they would change their views if forced to live with such suffering on a daily basis, and to weigh the affliction of genetic disease against the joy of parenthood, is also an open question. Science per se has little to contribute to such intensely personal judgments.

Though preimplantation diagnosis is controversial in some circles, to many observers it is both a major biomedical advance and one of the more benign procedures to emerge from the reading of human genes. Its application so far is restricted to a small, well-defined population: couples who know that they are carriers of an inherited disease whose genetic basis has been identified. The procedure's goal is humanitarian and economically sensible: enabling such couples to conceive children while avoiding the trauma and financial burden of lethal hereditary ailments.

Nevertheless, the enabling technology for this technique—the identification of disease-causing alleles and the development of tests to identify them—has other applications that may or may not be so benign. Tests for a disease-causing gene, for example, can be developed almost immediately after the gene is identified. But years, even decades, may elapse between the discovery of a gene and the development of therapies and cures for the disease it causes. During the uneasy period between discovery and cure, genetic testing can either reassure people at risk for inherited ailments or confront them with the challenge of learning more about themselves than they care to know. Not everyone is eager to consult a genetic crystal ball whose forecasts are both more reliable and more ominous than those of a carnival fortuneteller.

Recently, for example, a defective gene on chromosome 21 was identified in some families who experience a form of early-onset Alzheimer's disease known as familial Alzheimer's (FA). Healthy young adults in those families who are at risk for FA can now decide whether to take a blood test that will tell them if, at the prime of their lives, they will begin to lose their memory and ability to take care of themselves.

One person facing that decision is thirty-nine-year-old Debi, a kindergarten teacher with a hectic and demanding daily schedule. Matching her bright, cheerful classroom with a sunny disposition, Debi pays special attention to children with fractured home lives. She starts off each day by asking, "Who needs a hug?" As she rehearses her eager and energetic pupils for a Christmas pageant, it is easy to understand how pressures of the holiday season, not to mention the demands of juggling a career and three children of her own, could make her forget things occasionally.

But Debi, who watched her mother die of familial Alzheimer's, is at an age when every memory slip is cause for concern. Is her occasional forgetfulness due to stress, or is it the first sign of the dreaded disease? Debi knows all too well what the disease does to its victims. Although her mother was still in early middle age the last time Debi saw her alive, she looked much older, and could neither care for herself nor recognize her daughter. "She was a very, very bright person," Debi recalls, her eyes filling with tears at the memory, "and it was real hard to see her deteriorate."

Not surprisingly, Debi is terrified at the prospect of slowly, inexorably, losing her mind. "You really lose control of who you are and what you are doing, and that is humiliating, to be out of control. I think that is the hardest part. If I have the disease, I won't be me anymore; it really changes you a lot." Despite—or perhaps because of—these strong feelings, Debi is adamant about not taking the FA test. "If it was a bad answer, it would be too depressing. At least now I have hope that I am not going to be sick. I can live with the hope. I just don't want that burden."

Debi's sixteen-year-old son, Josh, is upset with her for not taking the test. Mother and son are very close, and Josh is tormented by uncertainty about her health. "She told me that the disease is worse than you can ever imagine," he exclaims, obviously perturbed. "I personally think imagining it is more horrible than watching somebody have the disease. Just the worry would make the rest of my life a living hell until she dies, naturally or from the disease." Once, in a rage, Josh told his school guidance counselor that, one way or another, he would find out if his mother had the gene, but he has become resigned to the fact that taking the test is his mother's decision. Still, fear of FA is an ever-present force in his life.

"The thing that I worry most about in my mom's behavior," Josh observes, working hard to maintain his composure, "is that she's having subtle losses of memory. She talked to a doctor, and the doctor told her it was stress. She has been under a lot of stress lately, which he said would cause some memory lapses. But you can't help but wonder if the disease is kicking in and taking effect, right under our noses."

Debi knows all too well what is going on in her son's mind. "I feel as if they're watching me," she says, "and waiting for me to do something really stupid. And they're gonna say 'Oops . . .' The reason I think that is because that's what I did with my mother. I was always watching her. And I know now how awful that must have been for her, to know that she was losing control and having everybody watch her do it."

The stress of watching and being watched weighs heavily on Debi's entire family. Josh admits that he has considered running away when the pressure gets too great. But Debi's cousin David can't run away from FA. He began to show signs of the disease at age thirty-eight and has deteriorated noticeably over the past five years. Talking to his wife, Cindy, about events of the past few days, the earnest, bearded man acts like an anxious child who knows he is disappointing his mother. He cannot recall what he ate the previous night during a dinner with his parents and has forgotten the name of the game he played with his nine-year-old son, David Michael. His speech is slow and halting, punctuated by so many pauses and fumblings for words that Cindy constantly guesses what he is trying to say. David, always eager to please, repeats her guesses, even if they are wrong.

Despite his obvious difficulty in communicating and concentrating on a single thought, this vigorous, athletic man understands what is happening to him. "I was a real smart person at one time," he says. "And then gradually I started saying . . . things seemed going . . . you know . . . I never remember. I remember everything and then I don't . . . now. Period. Sometimes, I can get it and other times it is just gone. I know what I am saying, but nobody else can understand me. Cindy can't even understand me sometimes."

Cindy holds David's hand and lovingly reassures him that he is doing very well, but she is frank about the disease. "You can see how he has difficulty in expressing himself," she continues. "There will be a point that he won't be able to really talk and move around and do things. He is a full-grown, adult man. I won't be able to get him in and out of the shower, if that is what he needs help doing. So there will be a time when he will have to go to a nursing home. Hopefully, that is down the road."

Getting Alzheimer's disease did not come as a complete surprise to David. Enough relatives had been afflicted, including his mother and maternal grandfather, that the household sensed it was a family problem by the time he was born. Two of his mother's siblings, Paul and Dan, watched a brother, their father, two sisters, and a nephew succumb to FA. Having reached their mid-sixties, Paul and Dan can be reasonably certain that they didn't inherit the deadly gene, which has struck its victims in their family during their late thirties and early forties.

Because Dan and Paul do not carry the FA gene, they needn't worry about their children and grandchildren inheriting it. But David and Cindy do have to worry, both about the way David's illness is affecting young David Michael, and about the boy's own medical future. Because a single copy of the FA allele is sufficient to produce symptoms of that disease, David Michael, like any child of an FA victim, has a 50 percent chance of inheriting the disease.

Cindy hasn't yet told her son that his father's illness is genetic. "There will be a point in time when I will explain that to him, but I don't think it is appropriate to worry a nine-year-old child about such a thing. There is no reason for him to worry about things that may never happen. My husband was so much older when this happened. By the time they realized what it was, the fact it was genetic, he was much older and much more able to cope with that. And I can still see how it affected him."

David Michael will be legally eligible to take the test when he reaches the age of eighteen. But at that point, will he want to? Will he choose the path that Debi has taken, preferring hope over certainty and avoiding the possibility of imminent bad news? Or will he hope, as researchers do, that the intervening thirty-year period before the disease is likely to manifest itself will see the development of a cure?

In the meantime, what about other people in similar positions? Most Alzheimer's cases are *not* caused by the particular gene identified in Debi and David's family. But a flurry of studies whose results were announced in 1992 and 1993 implicate at least two genes—one on chromosome 14 and one on chromosome 21—with cases of more common early-onset forms of the disease. The chromosome 21 gene has been positively identified; it codes for a protein called amyloid precursor protein or APP, abnormalities which have long been linked to Alzheimer's. The gene on chromosome 14 has not yet been identified, but seems to be located near a well-known chromosomal marker. And a third gene, identified on chromosome 19, is the first to be tentatively linked with the late-onset form of the disease.

In addition, there are hundreds of thousands of people at risk for other genetic diseases for which genes have been found and tests are available. The information provided by those tests can be a blessing, or it can cause a great deal of trouble, depending on how the tests are carried out and how the information is used. Cystic fibrosis, for example, is sufficiently common in the United States that some geneticists suggest population-wide genetic screening to offer unwitting carriers the chance to make informed reproductive decisions.[3] Tessie and Kevin Ashe discovered their carrier status through the worst possible route—having a CF child. In the wake of that discovery, preimplantation diagnosis allows them to make the best of a bad situation. But had they known of their carrier status before they were married, they could have considered several other options. They could conceivably have chosen not to marry, they could have decided at the time of marriage to adopt rather than conceive children of their own, or they might have contemplated artificial insemination with sperm from a noncarrier.

But ethicists remind us that the history of widespread screening programs is mixed and warn that great care must be taken to ensure that such programs are carried out in a compassionate and intelligent manner. In the early 1970s, for example, compulsory screening for sickle-cell trait was instituted in at least twelve states.[4] Screening was usually restricted to African-Americans (although 7 percent of subjects with sickle-cell trait were not black), and it was often conducted without informed consent. Inadequate regard for confidentiality, coupled with a preexisting history of racial prejudice, led to denial of insurance coverage and loss of employment for black people, including those who were simply carriers of the sickle-cell trait and showed no symptoms of disease. In addition, lack of adequate pretest education and posttest counseling caused intense psychosocial problems among many of those tested.

Properly designed and implemented screening procedures, however, can have positive results, as evidenced by a program aimed at Tay-Sachs disease. As many as one in twenty-five Jews of Eastern European descent are carriers of this disorder, characterized by nervous system degeneration, uncontrollable convulsions, total loss of sensory input, and death, usually by age four. After a nongenetic test for carriers was developed in the 1970s, the Jewish community mobilized to implement screening among its members. Combining pretest counseling and education with posttest follow-up counseling and literature, the screening program succeeded in lowering the incidence of the disease by 90 percent between 1970 and 1992. A similarly well-designed program, which targeted sickle-cell trait in Cuba, managed to lower the number of newborns with sickle-cell disease by 30 percent in 1989.

But these techniques and programs are just the beginning. With 4,000-odd genetic diseases already recognized, and with the Human Genome Project aiming to identify all of our estimated 100,000 genes, prospective parents will soon face decisions we can scarcely imagine today. Whether "bad news" from genetic test results leads to preimplantation diagnosis or abortion, a line must somehow be drawn between children who should be born and those who should not. How and by whom are such decisions to be made? Will parents be allowed to decide on their own, or will governments or insurance companies set limits on what couples can choose?

These are not abstract issues. One couple who knew that they were CF carriers asked their HMO to pay for a genetic test on their unborn child during pregnancy. They were told that if the child tested positive, the HMO would cancel their health plan unless they agreed to an abortion. Only threatened legal action induced the HMO to withdraw that demand. One can imagine the outcome of a similar situation involving a couple lacking either the financial resources or the determination to threaten a discrimination suit.

Current and future problems such as these are major concerns of Jeremy Rifkin and his Foundation for Economic Trends. The need for wider public understanding of relevant issues is particularly acute with human genetic testing, Rifkin argues, because many serious consequences won't be fully felt until decades into the future, long after legal precedents to govern the uses of those technologies are firmly in place. Rifkin is particularly concerned with the apparent convergence of genetic screening and genetic engineering. "Genetic engineering is the application of engineering principles to genetics," he notes. "What are engineering principles? Quality control. Predictability of outcome. Efficiency. Do parents continue in the time-honored tradition where children are a gift, a blessing bestowed by the Creator or some outside force? Or do parents see themselves now as architects, as designers, as programmers, for the next generation?"

Soon, Rifkin warns, there could be a powerful social message, "saying to parents: 'You have a responsibility not to burden us with children who may be 'defective,' who could increase the cost to society for their health and well-being.' So this changes fundamentally the relationship between parent and child. The child becomes increasingly a product that is programmed in advance. For parents, this kind of foresight is going to be terrifying. They are going to have to make decisions in advance of a child's coming into the world that could lock that child in for thirty, forty, fifty, or sixty years. We are not prepared to deal with those kinds of questions."

In the United States, where access to health insurance is an issue of growing concern, more than societal pressure is at stake. Information from genetic screening will soon enable testing not only for dozens of alleles that almost always cause serious disease, but also for many more alleles that simply predispose their bearers to certain conditions. Genes influence, to a greater or lesser degree, predisposition toward breast cancer, heart disease, diabetes, and multiple sclerosis, susceptibility to schistosomiasis and other infectious diseases, and scores of other ailments, both trivial and lethal. As researchers stake out more and more genetic markers and pinpoint more and more mutant alleles, it will become easier to distinguish, even before conception, people who are sick and likely to get sick from people who are well and likely to stay well. As a result, we are faced with a major breakup of what insurers term the "risk pool," the large groups of people among whom health care costs are shared.

If insurers and employers are allowed to require such tests as preconditions for insurance or employment, many will do so. The result could be catastrophic. "Increasingly in the 1990s," Rifkin predicts, "we are going to see a virulent new form of prejudice and discrimination emerge, based on one's genetic type. This is going to be a much more dangerous form of discrimination than skin color, than religion, than ethnicity." The easier and more widespread genetic testing becomes, ethicists warn, the more likely it will be for healthy people to ask, "Why should I have to pay for sick people? I'm not sick, and what's more, my family genetic structure is such that we're not likely to get sick."

Such discrimination could lead, some health care advocates argue, to the creation of a genetically based caste system in which the only people considered insurable would be those unlikely to suffer from serious disease, and the only people considered employable would be those unlikely to miss work or incur large medical bills. That "Brave New World" would enable new standards of industrial efficiency and cost-effectiveness, but it would do so by obliterating what we now view as equal opportunity for employment.

Is there any basis for these dire scenarios to suggest how people might respond to information of the sort provided by genetic testing? Back in the middle 1980s, observers predicted that policies concerning the test for human immunodeficiency virus would serve as de facto legal precedents for genetic testing.[5] Though HIV, unlike genetic disease, can be transferred from one person to another, once the virus has inserted itself into the host's genome it literally becomes part of his or her genetic endowment.

During that time, people at risk for HIV infection faced decisions about being tested that were much like those facing people at risk for FA or Huntington's disease today. Individuals who tested negative received a reprieve. Those who tested positive had no medically constructive course of action and risked losing jobs and health insurance. As a result, most people's responses were much like Debi's response to the FA test. Relatively few asymptomatic persons elected to take the test.

If HIV testing policies represent legal precedents for future genetic testing, there is indeed cause for concern. As soon as tests were made available, employers and insurance companies began advocating their use. Those actions were not based on fears about workplace transmission, but rather on a combination of prejudice, concern about job absences, and apprehension about medical costs. The early 1990s saw an explosion of lawsuits around AIDS-related insurance and employment discrimination. In one case, a Texas employer dropped coverage for HIV-related problems from $1 million to $5,000 after one employee was diagnosed, but maintained coverage for employees with other long-term, debilitating conditions. "Employers see cancer or heart disease as not the person's fault," Harvard School of Public Health professor Lawrence Gostin explained to the *Boston Globe,* "but see AIDS as the person's fault. They are making these moral decisions, which are unconscionable."

Those sorts of judgments are largely restricted to HIV testing today. But what happens when technology enables parents to prevent children from being born with genetic predispositions toward a host of conditions? Do children's genetic diseases then become the "fault" of parents' unwillingness to undergo or inability to afford various procedures? If so, when future parents face genetic testing, how will they feel obliged to act? Should they prevent or avoid the birth of a child with Tay-Sachs disease who will die within four years? How about a child with cystic fibrosis, likely to die in her twenties after much suffering? Or a child who would remain normal until age forty, but then be at higher-than-average risk for heart disease or breast cancer? Or a child who would be congenitally deaf but otherwise healthy? Or, for that matter, a boy who would mature at an adult height of five foot four rather than six foot two? Or is likely to be overweight?

The slope is slippery indeed. This is one situation in which many ethicists, and even scientists, tend to agree with Jeremy Rifkin's comment that such decisions are "beyond the ability of any parent in the world." The point is not to prevent the development of enabling genetic technologies, but to ensure that public awareness and dialogue create the proper social and legal settings for practices with which we can all live comfortably.

• • •

WHEN IT IS FINALLY possible to cure genetic diseases—not just alleviate their symptoms with drugs, but actually *cure* them by repairing or replacing defective genes—scores of researchers will share the credit. But when that day comes, there will also be one man whose dedication and vision will have been instrumental in launching the era of gene therapy. He is W. French Anderson, a researcher at the National Institutes of Health (NIH) who has dreamed of the genetic revolution in medicine for decades.

Anderson first resolved to "understand disease at the molecular level" in 1953, the year James Watson and Francis Crick discovered DNA's helical structure. No one had even found a gene, much less used one to cure anything. As early as 1968, Anderson discussed gene therapy in a journal article, to the surprise of his scientific peers. But Anderson is nothing if not single-minded; in 1984, he combed the minds of colleagues for a review article in *Science* that foreshadowed the revolution in gene therapy.[6]

On September 14, 1990, Anderson's dream became a reality, as a shy four-year-old girl received her first infusion of blood cells supplemented with a single, crucial gene of which fortune had deprived her at birth. Yet this first clinical trial of gene therapy was not a scientific breakthrough. "The science had been done several years before," Anderson asserts. "It was a cultural breakthrough, meaning an event that changes the way that we as a society think about ourselves. I don't want to take excessive credit, but basically it was something I had engineered and worked toward and built. My secretary," he concludes with a laugh, "said it just goes to show that I haven't had a new idea in twenty years!"

Friends say that Anderson thrives on challenge and competition. He seems to agree. "I have lived my life for the last few years walking along a cliff edge, with one foot over the edge and the other foot on a banana peel," he jokes. As gene therapy's most vocal proponent, he is a magnet for controversy—and a target for gene therapy opponents. "Genetic engineering is not the most popular profession," Anderson admits. "Bacterial genetic engineers have been firebombed in Germany. I have never had a personal threat, but I certainly know people who have. In fact, back in the 1970s, professors at Stanford were getting bricks thrown through their windows because they were working with recombinant DNA."

Fear for personal safety led Anderson to a fourth-degree black belt in Tae Kwon Do, a martial art he teaches at NIH. He also took an antiterrorist course and learned defensive driving skills by practicing on a race track. Race car driving caught his fancy, in part, he explains, because it "really puts you right on the edge. You take the car right to the edge of what it can do, so the tires are squealing just right. All four are gripping, and one mile per hour more and you will start to skid. But you hold it right at that edge."

This approach to life has helped place Anderson in the forefront of the gene therapy movement. Always pushing to the edge on curves, he has dedicated himself to making certain that when the techniques to perform gene therapy are developed, society will be ready to apply them. The first clinical trial of gene therapy illustrates how Anderson combines science with intense lobbying efforts. While working on his controversial gene therapy review article, he learned that adenosine deaminase deficiency (ADA), a deadly immune disorder, was a good candidate for gene therapy. As its name implies, ADA is caused by the lack of a crucial enzyme, adenosine deaminase, which normally protects the immune system's white blood cells from toxic substances. In its absence, the immune system is completely nonfunctional, and the mildest infections can be fatal. David, the Texas "bubble boy" who was forced to live his short life enshrouded in a plastic bubble to keep out all potential pathogenic organisms, suffered from ADA.

ADA is a rare disease; there are only about twenty-five cases in the world at any one time. But it is severe. Without therapy, affected babies usually die at a very early age, and nongenetic therapies are seldom effective for long. A few patients have been cured through bone marrow transplants from perfectly matched sibling donors, but these were exceptional cases. Most such operations failed. In 1986, scientists developed a treatment called PEG-ADA, which involves an enzyme supplement that works well for some ADA patients, but not all. Therefore a finite subset of ADA sufferers had little to lose by being genetic pioneers. Recognizing the potential of this situation, Anderson teamed with Michael Blaese, an ADA researcher at the National Cancer Institute whom Anderson had met earlier, to develop the therapy.

ADA deficiency is caused by a single, malfunctioning gene. Theoretically, if a normal copy of that gene could be placed into the patient's T-cells and induced to release its product, the disease could be treated. Doing that with single cells isn't really a problem; as you learned in the last chapter, mammalian cells can be transformed by injecting DNA into their nuclei. But the laborious process of manually treating one cell at a time couldn't work for ADA. To restore a patient's immune function, the gene would have to be injected into millions of circulating blood cells, clearly an impossible task.

Or was it? Scientists knew of creatures that successfully inject genes into many cells all the time: retroviruses. Certain genes carried by these viruses encode instructions for copying other viral genes, inserting them into human chromosomes, and directing them to produce new viral proteins. Watching this process in action, genetic engineers were both impressed and baffled by the virus's skill at manipulating our cells' molecular machinery. The researchers still haven't figured out just how the trick is accomplished; they don't know how to "write" the sort of genetic instructions viruses use to accomplish their ends. But several research teams realized that they could probably co-opt that viral intelligence, even if they didn't completely understand it.

They proceeded to "tame" the virus by using techniques similar to those discussed in Chapter 6—selecting, cutting, and pasting together stretches of genetic material from different sources. First, they chose a suitable retrovirus (*not* HIV), and removed sections of its genome until the resulting virus was crippled and could no longer multiply by itself. They then spliced the ADA gene into the remaining viral genetic material and added a gene from another virus to help "package" the molecular instructions inside a protein shell. When this artificial virus infected cells, instead of directing the production of new viruses, it would cause the production of the lifesaving enzyme.

The next step was to determine which of the patients' cells should receive that vital molecular cargo. At first, Anderson agreed with the prevailing judgment at the time: the T-cells in which the gene was actually needed were too short-lived to provide a lasting cure. For that reason, they resolved to target stem cells in bone marrow that live for a long time and steadily produce new white cells. Anderson and Blaese knew that if they could only insert the ADA gene into stem cells, they had a good chance of ensuring an adequate supply of ADA for life. Because doctors already knew how to remove and replace human bone marrow, prospects looked good—in theory. But preliminary experiments in monkeys revealed that in practice, bone marrow stem cells were hard to isolate and identify. The team couldn't find and genetically alter enough of them to produce a steady supply of ADA. The research was stalled for years.

Blaese broke the deadlock by persuading Anderson to concentrate on white cells instead. After all, he reasoned, some immune responses last for decades, so *some* T-cells must survive that long. In fact, other researchers were already using such cells to treat cancer patients. "I was so frustrated," Blaese recalls, "because we could cure these damn cells in tissue culture and we couldn't apply it to the kids. There was a good rationale for treating ADA deficiency with T-cells." Following that line of reasoning, Blaese, Anderson, and pediatrician Kenneth Culver together wrote the proposal for the first gene therapy trial.

That proposal had to face a special governmental review process on the safety of proposed gene therapy treatments, a process that Anderson himself had helped create. "Back in the early eighties," Anderson explains, "a number of people, including myself, felt that there had to be a public review, and we got the whole mechanism set up. Then, in May of 1986, when it became clear that it was likely we would have the initial protocol, I resigned, went to the other side of the table, and started to do battle with the process I set up in the first place."

The proposal was met with extreme criticism, even scorn. "There aren't too many names that people call others in polite company that I wasn't called," recalls Anderson. "And some that aren't used in polite company. The basic term was that this was premature, this was off the wall, this was just headline grabbing. This was leading to false hopes. Gene therapy couldn't be done for another ten or fifteen years. The whole scientific community seemed to feel that way."

Anderson attributes much of the criticism to the divergent views of research scientists and medical doctors. Basic researchers, he argues, look at this type of complex procedure and, by their very nature, pinpoint all the pieces of the puzzle that are not yet understood. They then suggest additional experiments needed to fill in existing knowledge gaps. Anderson and his clinically oriented colleagues, on the other hand, "are on the ward talking to patients who are dying *now*, not when it is a convenient time. Clinical doctors *know* that we don't understand the human body. We don't understand *most* of the treatments we use. The idea of understanding a treatment before you use it is almost not relevant. What is important is does it help? If it helps, you do it. If the risk is less than the benefit, you do it."

Anderson's critics, who included both gene therapy detractors and supporters, claimed that the team was rushing in without enough information. The Gene Therapy Subcommittee of NIH's Recombinant DNA Advisory Committee (RAC) wanted proof that the therapy would work better than PEG-ADA treatment. But because there was no animal model for ADA, the treatment's efficacy couldn't be proved before being tested in a human being. "Some people felt that maybe we shouldn't be taking this experimental step when there was already something that was working, at least partially," recalls Blaese. "The reason I felt we should go further is that although PEG-ADA improved the immunity in these children, it brought them into a midrange immune deficiency. Those children continued to have problems throughout their lives. Many of them don't live into adult life."

Yet as MIT's Richard Mulligan pointed out to the subcommittee, because the therapy did not target stem cells, it could not be a real cure. Indeed, Anderson planned to provide genetically altered cells every three months. There were also worries that the retrovirus might be dangerous: by inserting itself randomly into the cells' chromosomes, it could possibly disable a tumor suppressor gene or turn on an oncogene and cause cancer. This last concern worried Anderson enough that he, Blaese, and colleague Steven Rosenberg of the National Cancer Institute performed a different gene therapy experiment first: they put genetically altered white blood cells into terminally ill adult cancer patients. The gene they put in would not help these patients—it was only a "marker" so that they could follow the cells in the body. But the experiment convinced Anderson that using retroviruses was safe enough to try on young children.

Not everyone agreed. Jeremy Rifkin's group filed suit in federal court in Washington to try to stop the gene therapy experiments, criticizing the NIH for not opening final hearings on the experiment to the public. "No scientist at the cutting edge of molecular biology wants to ponder and reflect about the potential consequences of the research they are doing," Rifkin argued. "It just slows them down. So the whole idea that the scientists and the corporations involved in this research are going to self-police, that they are going to ponder, be reflective, pause, is just naive thinking."

Rifkin's legal efforts delayed the gene therapy experiments for a time and produced a settlement in which NIH agreed to open hearings to the public. On September 14, 1990, at 8:59 A.M., after eight months of extensive reviews by seven different boards, final approval for the ADA trial was given. At 12:52 P.M., genetically altered cells quietly started dripping into the veins of a four-year-old patient, Ashanti, who played on her hospital bed during the historic procedure.[7] Anderson was much more nervous than his young patient, for he knew that the fate of human gene therapy rested on the outcome of this trial. "Everything went on line for this protocol. If it had failed, I would have been ruined, and it was all over. But if it succeeded, everything was primed to go."

Luckily for Anderson, Ashanti, and proponents of gene therapy, the ADA protocol, though by no means a perfect or final solution to the disease, seems to have succeeded. For the first time in Ashanti's life, she can mount an immune response to foreign substances. As a result, she can attend kindergarten, something her parents had only dreamed about.

The De Silvas, after spending years living in constant fear that an ordinary childhood illness such as chicken pox would end Ashanti's life, are delighted with their child's newfound resilience. "When the whole family fell ill with influenza," her father, Raj De Silva bubbles, "the first to recover was Ashanti! And this runny nose she had all the time disappeared. After that we began to feel comfortable and gradually began to relax even without knowing it. So much so that we started planning to send her for swimming lessons!"

Anderson views the apparent success of the ADA trial as just a beginning rather than an end in itself. For his dreams go light-years beyond that protocol, to a day when patients can walk into a clinic and receive an injection of gene-carrying viruses aimed specifically at the cells whose defective genes cause disease. He knows that researchers must learn a great deal more about viral molecular intelligence before they will be able to design such sophisticated gene delivery systems.

The path before them, however, is clear. "A hepatitis virus targets the liver; mumps, the salivary glands; rabies, the nervous system," Anderson points out. "So the ideal is to learn how viruses do it and then build engineered virus-type structures that will do what we want—and go to the right organ. We're sort of mechanics right now, not engineers," he continues. "You just sort of plunk this together and plunk that together and put it in. Now it's time to learn how to engineer and to build precisely. So we're horse and cart, plunking around. What I want to do is build an automobile."

• • •

TWO EXAMPLES of other gene therapies under development show that the ADA trial was not a fluke, yet remind us of just how far we have to go before achieving Anderson's dream. The first case concerns a disease involving the liver, and the second returns us to the lungs, where the sort of engineered viruses Anderson dreams of may offer a cure for cystic fibrosis.

Familial hypercholesterolemia (FH) is a hereditary ailment that afflicts nearly one in 500 people in the United States. Unlike cystic fibrosis and sickle-cell anemia, two copies of the defective gene aren't necessary to produce significant symptoms of this condition. The affected gene produces a cellular receptor that detects low-density lipoprotein (LDL) cholesterol—the "bad" cholesterol that causes atherosclerosis—in the bloodstream and enables liver cells to remove any excess. People who carry a single defective copy of the LDL-receptor gene have cholesterol levels two to four times higher than normal and face a high risk of atherosclerosis as early as their middle thirties. Their prognosis can be improved by combining dietary restrictions, exercise, and cholesterol-lowering drugs. *Homozygotes*, however, have no functioning LDL receptors, cannot remove cholesterol from their blood, and have cholesterol levels six times normal or higher. Even rigidly controlled diet and drug therapy offers little help in this situation; homozygotes may have heart attacks before age ten. Double heart/liver transplants have been attempted and proved that receipt of a normally functioning liver could lower a patient's cholesterol dramatically. But organ rejection problems have so far doomed transplant recipients.

James Wilson at the University of Michigan was inspired by the heroic efforts to save FH homozygotes, but realized that organ transplants weren't the best solution. "The more elegant approach," Wilson suggested, "would be to do gene surgery, so to speak, rather than organ surgery." With luck, he reasoned, gene therapy could restore enough cholesterol-lowering function to homozygotes' livers to allow drug and diet treatment to work. It was certainly worth a try. Drawing from Anderson's experience with the RAC committee, Wilson realized that his bid for approval would be easier if he could find an animal model in which to prove the effectiveness of his approach before testing it in humans. Luckily, such a model for FH did exist: a laboratory strain of rabbits displays a genetic disorder that closely mimics FH in humans.

Beginning with those rabbits, Wilson developed a gene therapy protocol similar to that used in ADA therapy. First, his team engineered a crippled retrovirus to carry the FH gene into liver cells removed surgically from the subject. Those transformed cells are then returned to the liver via a tube leading directly into the large vein which feeds that organ. The real question was whether the engineered cells would survive in the body and effectively clear cholesterol from the bloodstream.

Although Wilson estimates that he successfully transformed only 4 percent of the experimental rabbits' liver cells, their cholesterol levels fell 30 percent—from 600 milligrams per deciliter of blood to about 425. Although this is still far above the 90 milligrams per deciliter found in normal rabbits' blood, the results were encouraging enough for Wilson to push ahead to the next step—testing the protocol on an animal more closely related to humans. That work was performed on baboons, which passed the safety test with flying colors. Wilson could introduce normal human cholesterol receptors into liver cells with no adverse effects.

Many questions still remain. Will the limited number of transformed cells limit the procedure's effectiveness? How long will the transformed cells live? Liver cells normally live for a fairly long time, and animal studies show that genetically engineered cells are stable for at least six months. Whether results in humans will be that good or better is hard to predict.

The positive signs are good enough to merit clinical trials in humans. Because the technique is so new, it can be tested only on patients already hampered by advanced coronary artery disease, people too far along for the procedure ever to make them truly "normal" again. Still, Wilson is encouraged by his new ability to offer potential alternatives to inevitably progressive disease, permanent disability, and early death. He hopes in the not-too-far-distant future to develop a gene delivery system that will carry the LDL-receptor gene directly to the liver cells, eliminating the need for surgery altogether.

ANOTHER RESEARCHER working feverishly to achieve Anderson's dreamed-of proficiency in genetic engineering is Ron Crystal, chief of the Division of Pulmonary and Critical Care Medicine at New York Hospital. Crystal is convinced that gene therapy is the only way to treat cystic fibrosis patients, but he can't use the approach employed on ADA and FH because he can't simply pluck cells out of his patients' lungs for treatment. What's more, lung cells divide very infrequently, and retroviruses can incorporate their genes only into cells that divide. Delivering the CF gene to the right place therefore requires a different sort of gene delivery procedure.

A possible solution to the problem occurred to Crystal soon after he spoke with a French colleague working with adenoviruses. While Crystal was jogging the next day, inspiration struck. "I don't know why I thought about it, because my laboratory was not working on cystic fibrosis. In fact, the cystic fibrosis gene had not yet been found. But suddenly I got this idea. The adenovirus loves the lung—you get colds and respiratory infections from it. And we realized that by using an adenovirus as a vector, we might be able to modify it with respect to the cystic fibrosis gene in such a way as to transfer it into the cells lining the lung in a living person—and essentially cure the respiratory manifestations of cystic fibrosis."

The more Crystal researched adenoviruses, the more perfect they seemed for curing cystic fibrosis. Unlike retroviruses, they didn't require cell division to work, so they could affect even nondividing lung cells. Engineered viruses could be packed very densely into the airways in the form of an aerosol, giving at least some of them a good chance to reach their target cells before the body's defense mechanisms destroyed them all. And the adenovirus inserted genes into the cell and made them work without splicing them into the chromosomes, thereby eliminating fears that a wayward virus could cause cancer.

Crystal's colleagues first proved that adenoviruses in laboratory culture can "fix" defective cells from the pancreas and lungs of a CF patient. Then Crystal tested the specificity of the adenovirus on the lungs of rats that were as susceptible to adenovirus as humans are. Because no animal model for CF exists, Crystal used marker genes to track the uptake of the adenovirus into the rat's lung cells. Soon after the rat trials produced positive results, the CF gene was found, and further trials carried that gene into the lungs of primates to continue testing the treatment. All went well, and in December of 1992 Crystal obtained approval from RAC. The FDA approval followed on April 16, 1993. The first patient was treated on April 17, the second patient one week later.

Crystal was always optimistic about the outcome. "I'm sure we can convert the cells in the lungs from abnormal to corrected," he stated confidently as trials began. "No question at all. We could do it. The question is, Is it *safe* to do it?" For this reason, during the first human trials Crystal inserted a tube into the patient's lungs and dripped the genetically altered virus into airways feeding only part of the lung. In this way the doctors could determine whether cells in this area responded to the treatment. The first people to receive this procedure were older CF patients—people whose lungs had already been badly scarred by the disease. Thus far, no ill effects have been reported, and the viral vector seems to be doing its job. If all continues to go well, a patient's entire lungs will be treated, perhaps by inhalation of the virus in aerosol form.

Anderson, for his part, is working on another targeted vector, one that can travel through the bloodstream and insert the gene coding for insulin—the enzyme that controls blood sugar levels in the body—into the pancreatic cells of diabetics who lack it. This is a much trickier goal than ADA and FH therapy, which have only to get their genes in place and turn them on. The production of the insulin protein from its gene must respond to blood sugar levels: the gene must be turned off when the concentration of glucose is low and on when it is high. Assembling a piece of DNA with all the right promoter regions to operate properly in a foreign environment would be a major achievement.

To assemble resources for this ambitious venture, Anderson joined forces with Genetic Therapy, Inc. (GTI), a biotech company. Though initially wary of formal research collaborations between basic researchers and industry, Anderson ultimately took advantage of a 1986 act of Congress allowing partnerships between NIH scientists and corporations. Anderson is quick to point out that he has no formal title at GTI, owns no stock in the company, and receives no money for the commercialization of his efforts. "I am very excited about the collaboration with GTI," he says. "The first gene therapy protocol took place at least two or three years earlier than it would have had we not had GTI's help."

Collaboration with industry is essential to Anderson's ultimate goal for gene therapy: use of genetic manipulation to attack every disease, not just those traditionally believed to be hereditary. "Every disease has a genetic component, whether it's propensity for breast cancer or heart disease or whatever," Anderson remarks. "So what you're talking about will be applicable to every disease—arthritis, Alzheimer's—every disease. But what makes gene therapy particularly exciting isn't curing them, it's preventing them in the first place. Because the Genome Project will show us the range of tumor suppressor genes and the range of atherosclerotic genes and so on. If a person has a propensity, for example, for breast cancer, you simply supply her with the correct version of the gene so she never gets cancer in the first place. That's where gene therapy is going."

But the low cost and broad application that gene therapy proponents hope for will bring its own problems. When replacing or eliminating a gene that causes cystic fibrosis or cancer is no harder than buying a pair of eyeglasses, people may not think twice about using gene therapy. What if people decide to use these procedures to "enhance" otherwise normal characteristics? As the Human Genome Project leads to the discovery of genes controlling eye color, skin color—possibly even intelligence and compassion—will people be tempted to change themselves at their most basic level? Will parents make sure that their children are born with certain characteristics they find superior?

The evidence indicates that people would at least be tempted to do so. Billions of dollars are spent each year on cosmetic surgery for strictly aesthetic reasons. Remember that some people are already seeking to administer growth hormone to make short children taller. As you might expect by now, critics like Jeremy Rifkin see more widespread implications. "Increasingly, we are going to see the disabled as defective products," he warns. "I can see a day coming very soon in the 1990s when people will look at someone walking down the street, a young person, and say, 'Why did that young person have a cleft palate? He must have come from a lower class. His parents couldn't afford to program that trait out at conception. Why is that person disabled? Obviously she comes from a class that couldn't afford genetic screening or genetic engineering.'"

Though it surprises some colleagues, French Anderson is an outspoken critic of enhancement engineering. "I'm absolutely opposed to gene transfer for anything except treatment of serious disease. The only way to prevent that is not by legislation, because that's not going to work. It's not by any sort of idealistic idea. It's the public awareness. If the public is aware of what is happening and what can happen, whether it's abortion or fetal tissue, whatever it is, the public is able to communicate its wishes. Might we be able to actually damage our species with indiscriminate use of gene transfer?" he asks. "Now I talk like Jeremy Rifkin! But basically, in private, Rifkin and I are in complete agreement on this particular subject. We simply disagree in public about how to present that information to the public."

• • •

THERE IS YET one more issue—perhaps the most philosophically fascinating and disturbing one—that is raised by human genetic engineering. As our expertise at genetic manipulation improves, why should we need to deal with genetic testing and preimplantation in each generation? All the techniques we've described so far concentrate on the *somatic cells* of the body, those which constitute lung or liver tissue but don't determine the genetic constitution of future generations. Why not actually fix a defective gene in the *germ-line cells* that produce sperm and eggs and permanently free an entire family from inherited disease?

Both supporters and critics of germ-line therapy agree that the prospect will be tempting to patients and families. "If somatic cell therapy works as people hope," says ethicist Leroy Walters, "there are going to be some families in which the grandfather was treated successfully, then perhaps a daughter and then perhaps a grandson. And pretty soon the family is going to ask itself, 'Wouldn't it be more efficient to do this repair once and for all? At least for our family?' Obviously the savings with germ-line therapy would be tremendous. Only one individual would have to be treated, and from then on the family would not have that problem of genetic disease."

Most people agree that decisions we make now should not irrevocably change our children. "We have no right to decide what future generations of humans should be like," says Mario Capecchi of the University of Utah. "So we should restrict our use of this technology to work on somatic tissues and thereby help only the individual and not affect future generations. I think that is where the biggest misuse is possible." The consensus on this issue to date has given rise to regulations requiring researchers submitting proposals for gene therapy to prove that the treatment will affect only somatic cells.

One of the problems is that there may be beneficial effects of what seem to be completely detrimental genes. The gene for Hemoglobin-S, once thought to be a strictly deleterious allele, is now known to protect *heterozygotes* against malaria. Other theories hold that genes for other diseases may have similar hidden benefits—benefits that we might someday learn to exploit without suffering the gene's negative effects. Some researchers hypothesize, for example, that carriers of Tay-Sachs disease may have benefited from some protection against tuberculosis, a deadly killer in the crowded Jewish ghettos of eastern Europe. "So if we choose to eliminate such genes from the population," says Rifkin, "we may be programming our own extinction in the long run."

Scientists assure us that the enabling technology for germ-line therapy is still in the future. But even now, a working group at RAC is considering the ethics of that procedure. The public may need to decide on these issues sooner than it cares to; in the past, technology has been developed much quicker than even the scientists involved could dream of. "I remember," says Ashanti's mother, "Dr. Sorenson took Ashanti's blood. And he told me, 'This is for possible gene therapy when your daughter is sixty years old.' I thought 'Oh, how wonderful!' but wished it could happen sooner. And it happened real fast, you know. She was only four."

Jeremy Rifkin has vowed to block any use of germ-line therapy. "Every decision by a scientist to edit millions of years of evolution is a eugenics decision," he argues. "Perhaps none of us are wise enough, have the clairvoyance, the wisdom, to dictate basic changes in millions of years of genetic evolution. I don't think any of us should have that power. I think it's an unwarranted power and should not be exercised."

"There has been a determination in the public that somatic cell gene therapy is good," says French Anderson. "Germ line—nobody knows what they think about that yet. But as the discussion takes place, society will come down on one side or the other. I think the side they'll come down on is just like the somatic cell: the use of germ-line therapy to treat serious disease is good; the use of it to treat any other purpose is bad, it's wrong. These are major considerations that really require and have had a lot of thought," says Anderson. "But the somatic-cell debate is over. The germ-line one has begun."

8

Genes, Mind, and Destiny

Our destiny exercises its influence over us even when,
as yet, we have not learned its nature.

Friedrich Nietzsche

The tissue of life to be
We weave with colors all our own,
And in the field of Destiny
We reap as we have sown.

John Greenleaf Whittier

OSKAR STOHR AND JACK YUFE, identical twins separated at birth, seem to have lived radically different lives; Stohr was raised Catholic in Germany, while Yufe grew up Jewish in Trinidad. They had met only once, when they were in their late twenties. Yet when the middle-aged twins met again—on the day they reported for participation in Thomas Bouchard's personality studies at the University of Minnesota—they looked and acted remarkably alike. According to a "Research News" article in *Science,* both men appeared wearing "blue double-breasted epauletted shirts, mustaches, and wire-rimmed glasses."[1] The article continued to describe the personality traits shared by the pair, including hasty tempers and odd senses of humor; both enjoy "surprising people by sneezing in elevators."

Another pair of identical twins in Bouchard's ongoing study, separated when they were five days old, both became volunteer firemen in New Jersey. When they met as adults, according to *U.S. News and World Report,* both Mark Newman and Gerald Levey sported "droopy mustaches, aviator-style eyeglasses, and a key ring on the belt on the right side." The magazine's brief portrait, titled "The Case of the Identical Firemen," also noted that both twins "love John Wayne movies, drink beer alike, and may carry a gene for risk taking."[2] Regarding their shared leaning toward volunteer firefighting, the brothers concur that "firemen aren't made, they're born."

This article typifies the penchant of the media—and some scientists—for drawing broad conclusions about genetic control of personality traits ranging from shyness, intelligence, and taste in clothes to preferences for particular types of jewelry. Emphasizing similarities between identical twins, who begin their lives with virtually identical genetic endowments, some stories go even further. The *U.S. News and World Report* piece asserted that "a growing number of scientists" believe that family environment and early experiences "may be largely incidental" to the kind of person a child will become, because infants are "irreversibly stamped" by their genes.[3]

Such glib pronouncements and the studies that support them are becoming increasingly popular. They make intuitive sense to parents watching two or more children develop distinct, yet overlapping, personalities and talents. They seem to be logical extensions of studies that describe how genes cause cancer, produce debilitating disease, or help children grow tall. Yet nearly every research report in human behavior genetics ignites fierce scientific debate. Why? Because the science behind theories of human behavioral genetics is complex and fraught with methodological problems, and because scientific opinion on the roles of "nature" (genes) and "nurture" (environment) in shaping human behavior has oscillated wildly over the years.

Around the turn of the century, most scientists and laypeople believed that humans differ in mental ability and personality primarily because of their "nature"—innate, biologically inherited differences. That extreme biological determinism fell from favor after World War II, partly because of revulsion at Nazi atrocities based on pseudoscientific notions of racial superiority. Over the next two decades, studies revealed that nature and nurture are partners rather than alternatives in shaping organisms' nervous systems. Hence, animals' behaviors are influenced both by genetic instructions and by interactions with their environment as they grow. It is an undeniable fact of biology that an individual's genome *affects* the construction and operation of his or her brain. But the extent to which specific DNA sequences *control* and *determine* human behavior remains an open question.

Uncertainty notwithstanding, the past twenty years have seen a revival of the false dichotomy between genes and environment and a swing in the pendulum of scientific opinion toward hereditarianism. Today there are ongoing searches for genes that cause alcoholism, manic depression, criminality, language proficiency, intelligence, and sexual orientation. In part, that shift is understandable. The tools of molecular biology have pinpointed genes at work in evolution and disease, so why not apply the same techniques to behavior?

Part of the controversy springs from the nature of the research itself. Human behavioral genetics deals with the two most complicated living systems known: the human genome, which consists of at least 100,000 genes, and the human brain, which contains more than 100 billion nerve cells. Understanding either genome or brain alone is a challenge. Predicting their interactions is staggeringly difficult. Recall how arduous it has been to associate individual genes with such simple, well-defined disease states as cystic fibrosis and ADA deficiency; it is much, much harder to draw causal links between specific alleles and complex, ill-defined traits as "intelligence" or "shyness."

But disputes in this field involve much more than just methodological difficulty. For the really important questions aren't whether Oskar Stohr and Jack Yufe enjoy sneezing in elevators because they share a stretch of DNA, or whether Mark Newman and Gerald Levey were truly "born" firemen. Those are trivial matters. The vital issue is whether specific, identifiable genes can make certain people or racial groups shy or rowdy, honest or sneaky, clever or stupid, alcoholic or manic-depressive, heterosexual or homosexual, and whether environmental influences can alter genetic "destiny."

That's where the controversy begins. Society has always distributed wealth, health, and power unevenly among individuals, sexes, and races. In the past, when the aristocracy ruled by divine right, socioreligious doctrine legitimated those inequities. Today scientists, rather than clerics, are called upon to explain why society works the way it does. Yet many scientists are—or profess to be—unaware of how easily scientific claims, even uncertain ones based on tentative and preliminary results, can be used to "explain," and hence justify, why some individuals or groups of people prosper while others languish.

"This," argues Tim Tully of Cold Spring Harbor Laboratory, "is the dark side of molecular biologists turning their attention to mechanisms of behavior. They don't understand either the social or biological environments into which they are dropping their results. And it is going to cause more damage than can be imagined."

• • •

To GARY GROSS, debates about genes and behavior are far from abstract. Gross has suffered most of his life from alcoholism, one of the conditions some researchers believe has a genetic basis. He is sober today, thanks to the strength of his own resolve and support from his wife, Kim Collins, and son, Ian. But for years Gross was so seriously addicted to alcohol that he not only placed himself and others at risk, but was blind to liquor's role in his predicament.

"I think I was in a bar in a college town," Gross recalls, recounting an episode of drunken driving, "and coming home, I woke up and saw a telephone pole swinging over my head. That is all I remember of that accident, which is as close as I probably ever came to death or serious injury. I clipped a telephone pole in half. Ruined my car. But even right after that, it never occurred to me, 'Hey, I had too much to drink.' Not once did I say to myself, 'Wow, stop drinking.' It was, 'Wow, I was tired, I fell asleep.'"

Alcohol continued to affect Gross's personality after his marriage. "He would have the first couple of beers," Kim Collins says grimly, "and it would be okay. After that his behavior would slowly change. He would get mean and belligerent, and I wouldn't like to be around him. He and I would have tremendous fights and I would beg him to stop drinking. Ian would say, 'Daddy'—because he was just little at the time—'Daddy, please don't drink.' When he was sober, I would try to talk to him and he would get angry and tell me that the problem was mine."

Ian himself has vivid memories of that period, memories that contrast starkly with the image projected by alcohol commercials on TV. "You know, 'It's Miller time!'" he observes sarcastically. "And you see all the guys having fun and stuff. But at my house when it was Miller time, my dad would get mad, hit my mom, scream, throw things, yell at me. That's the Miller time I am used to."

As is often the case with alcoholics, Gross's drinking didn't appear in a vaccum. Alcohol was readily available in his home. Given the prevalence of drinking in his family, and given what happened to Gary, it is hardly surprising that he and Kim Collins wonder about their son

Ian's future. Did Gary learn to drink by imitating his relatives? Or does he carry an alcoholism gene that he might have passed on to Ian, making him susceptible to the disease as well?

Gary has strong opinions on the matter. "I don't feel that I was born an alcoholic," he states firmly, arguing that he learned by imitating his father and friends in both social and business settings. Kim, on the other hand, fears that heredity is at work here. "Because Gary drank," she reasons, "because so many people in his family drank, it makes me believe that it is probably genetic."

Alcoholism clearly "runs in families," as the popular saying goes, but *why?* Scientists ask the same question—and come up with the same range of conflicting opinions. To help researchers test their hypotheses, the Gross family is participating in an ongoing study, run by psychiatrist Ernest Noble of the University of California at Los Angeles, which is one of several aimed at linking specific genes with various forms of alcoholism.

Noble's original work on this subject, carried out in collaboration with Kenneth Blum of the University of Texas Health Center at San Antonio, was published in 1990 in the *Journal of the American Medical Association.* In that paper, Noble and Blum explained how they examined the brains of thirty-five alcoholics and thirty-five nonalcoholics, searching for genes associated with alcoholism. Their results led them to single out a gene that codes for a particular nerve cell receptor protein. "There are two major categories," Blum explains, "genetic contribution, which we believe is 60 percent, and environmental contribution, which is 40 percent. When we look at the genetic contribution, we believe half of that comes through . . . this gene. It is the single major gene in alcoholism."[4]

The study, initially greeted with great enthusiasm by the scientific community and the popular press, was hailed as a major stride toward more successful research and treatment of mental illness. Noble suggested that his team's work would help develop a blood test to identify people susceptible to alcoholism, thus singling them out for preventive intervention. The research received front-page coverage, not only in scientific journals, but in the *New York Times,* the *Wall Street Journal,* and on numerous television news reports.

But almost before the ink on those news reports was dry, rebuttals appeared. Another *JAMA* paper, written by a team at the National Institute on Alcohol Abuse and Alcoholism, failed to reproduce Noble's results.[5] Still another study, published in the *Journal of Abnormal Psychology,* warned that researchers should not ignore "the significant influence that the environment has in the origins of alcoholism." Since that time, further studies by both sides in the debate have raised doubts, charges, and countercharges that continue to fly back and forth in both scientific journals and mainstream media. "The major problem," observes Paul Billings of Stanford University, "was that the scientists didn't suggest that this was a very preliminary finding and would need to be reproduced before announcing that they had found a 'gene for alcoholism.' It fit very much into a climate in the scientific world and in society at large that was searching for a biological quick fix to the problem of alcoholism."

Where does this research stand today? As is typical for human behavioral genetics, the answer depends largely on whom you ask. An opinion poll among scientists would find reasonable, though by no means unanimous, agreement that susceptibility to alcoholism is at least partially heritable. But consensus ends there. Some researchers believe in the existence of a single alcoholism gene, while others think several interacting genes lie behind the disorder. Still others, certain that many genes are involved in various aspects of the disorder, are equally certain that genetic tests for alcoholism-prone individuals will never become any simpler, more accurate, or more effective than the psychosocial tests presently in use.

What is the background for this fracas? Is there proof that any individual genes *control,* that is, *determine,* any human behaviors? To answer these questions, and to explore the scientific, social, and political arenas in which human behavioral genetics research must operate, we must first examine studies of gene-environment interaction in species more amenable to study than our own.

• • •

"BEHAVIOR" ACCORDING to *Merriam Webster's Collegiate Dictionary,* tenth edition, is simply "the manner of conducting oneself." A textbook definition is more precise: "Externally visible activity of an animal, in which a coordinated pattern of sensory, motor, and associative neural activity responds to changing conditions."[6] In plain English, the second definition points out that three processes underlie everything animals do. We sense events around us, evaluate that information, and respond to the situation.

You might wonder how genes could control these complex processes. The stretches of DNA we've met so far perform clearly delineated tasks: some code for proteins that build body tissue, some code for enzymes that control life processes, and some control groups of other genes. How could this precisely targeted biochemical intelligence govern something as intangible as the ability to remember a child's face, to respond to alcohol in a specific fashion, or to excel on IQ tests?

The basic answer is simple. The three components of behavior are produced by classes of specialized cells: sensory receptors, neurons, and muscles. These "information receivers," "decision makers," and "action enablers" are wired into networks whose complexity makes computers look like tin-can telephones. Where does the "wiring diagram" for those networks come from? You've seen in earlier chapters that genes guide the formation of every tissue in the body. So genes *must* influence the construction of the eyes, ears, brain, nerves, and muscles, hence the way they work. No amount of training can enable a chimpanzee to communicate with the skill of a ten-year-old human child; this disparity in intelligence exists thanks to differences in brain structure and function whose origins trace back to differences between human and chimpanzee genomes.

Beyond that general observation, many specific studies link heredity and animal behavior. Researchers have agreed for years that certain "instinctive behaviors" must be largely pre-programmed by genetic instructions because many animals perform such behaviors without prior experience. A female tick, for instance, responds to only a few sensory cues with a few predictable actions; she has no time to learn what to do and little room for error. Early in life, she aims for bright light by climbing upward on leaves and branches. Once perched, she waits—as long as twenty years—until she detects the scent of butyric acid, a component of sweat that signals a mammal passing beneath her. At that point she releases her perch and drops (if she is lucky) onto the skin of her prey. Responding to the warmth of blood near the skin surface, she burrows in and drinks any warm liquid she finds until she fills to capacity, drops to the ground, lays her eggs, and dies.

Many more elaborate behaviors seem to be pre-programmed as well, including the web-building behaviors of certain spiders and nest-building activities of certain birds. Another example is the egg-retrieving behavior of a greylag goose brooding her clutch. If such a female notices an egg nearby, she stands, reaches out her beak, rolls the egg gently into the nest, and settles down upon it. At first glance, that seems unremarkable. But if the egg is snatched away after the behavior begins, the goose goes through the motions nonetheless, carefully retrieving and incubating an egg that is no longer there! What's more, the same actions can be triggered by a variety of objects, including beer cans and baseballs, which resemble eggs in that they are convex and have rounded edges. Oddly enough, a separate behavioral program, which responds to a different set of sensory cues, guides the goose in removing empty eggshells and foreign objects from the nest. In that program, the carefully retrieved beer can is recognized as a "non-egg" and ejected.

Humans, too, exhibit innate behaviors. A newborn baby, held to a breast it has not yet touched by a mother it does not yet know, instinctively begins the coordinated muscular activity we call suckling. That behavior can also be triggered by baby bottles or pacifiers which exhibit certain features of a lactating breast.

Dogs provide a classic example of genetically influenced behavior, evidenced by animal breeders' success in shaping habits, including hunting, aggression, herding, affinity for horses, and loyalty in various breeds. Dog lovers' observations and scientific studies concur that different breeds exhibit an extraordinary range of behaviors and dispositions. Newfoundlands, for example, are water rescuers par excellence; they leap into the water and swim long distances to help drowning humans. Along with thick, oily coats and webbed feet, these animals have an instinctive affinity for water; a Newfoundland puppy presented with a large water dish is likely to climb into it. Border collies, on the other hand, which have been bred from ancestors with a history of herding reindeer, have an instinctive drive to gather things together; they do their best not only to herd sheep or cattle, but also groups of children and even tennis balls. Other distinctive behaviors include a tendency to crouch down flat and stare at animals in a herd and to avoid barking at their charges. Presented with water dishes, these dogs drink from rather than attempting to swim in them.[7]

Genetic studies of natural behavioral variants in other animals have produced fascinating if inconclusive results. Nearly sixty years ago, for example, the bee-keeping industry was ravaged by the bacterial disease American Foul Brood. Luckily, certain bee strains proved resistant to the plague, thanks in part to a specific series of "hygienic" behaviors that helped curtail its spread. Resistant colonies, it turned out, quickly located dead pupae in the honeycomb, uncapped their cells, and removed them from the hive.

Early breeding studies suggested that hygienic activity could be broken down into two components—"uncapping" brood cells and "removing" dead larvae—and that each of these behaviors was "switched on" by a single gene. Later experiments showed that the situation was more complicated; while most variation in hygienic behavior was accounted for by these two "master switch" genes, an undetermined number of other genes exert smaller but significant effects.

Most importantly, studies of species after species made it clear that normal behavioral traits can be influenced by genes, and that individuals within a species can exhibit inheritable variations in those traits. That variation is critical, because, as discussed in Chapter 2, the combination of inheritable variation and differential survival enables evolution to occur. Observations in comparative animal behavior have turned up powerful evidence that certain behaviors have evolved over time, just as body parts and cellular functions have.

Building on this foundation, Edward O. Wilson of Harvard and several colleagues founded the field of sociobiology, which they define as "the systematic study of the biological basis of all social behavior."[8] Wilson's magnum opus *Sociobiology* is an extraordinary treatise on the evolution of behaviors in animals ranging from slime molds through social insects—Wilson's personal forte—to mammals. That book might or might not have aroused much interest outside of scientific circles had it not concluded with a chapter that dared to "consider man in the free spirit of natural history, as though we were zoologists from another planet completing a catalogue of social species on Earth."[9]

Wilson and his colleagues have since developed and expanded the thesis that "a great deal of human behavior is affected by heredity, and hence can be explained more readily by biology than by the usual formulations of the social sciences."[10] Over time, sociobiologists have discussed, in largely theoretical terms, the notion of genes in humans and other animals that influence aggression, dominance, altruism, selfishness, and other similarly complex behaviors.

But while it is possible to argue that genes influence behavior in a general sense, it is quite another matter to show that a *specific* gene—or pair of genes, or even a score of genes—actually control *specific* details of an animal's responses to its environment. At this point, it is fair to ask whether anyone has found, in the strict molecular sense of locating and manipulating, any stretches of DNA that affect specific behaviors predictably.

This question leads us once again to *Drosophila*, the geneticists' old standby, partly because its chromosomes are so well known and partly because its nervous system is neither so complex as to be intractable nor so simple as to be uninteresting. Although those of us more accustomed to swatting flies than observing them may not think so, these insects do exhibit interesting behaviors.

Male fruit flies, for example, can't mate with whomever they choose; they must convince females that they are desirable partners. To do that, they sing courtship songs which, though neither audible nor attractive to humans, enchant female flies. According to behavior geneticist Charalambos Kyriacou of Leicester University, the love songs of several *Drosophila* species consist of two main components: a hum (mmmmmmm) alternating with several sharp pulses (zi..zi..zi..zi..zi). Perhaps to keep things from getting boring, the flies increase and decrease the amount of time between the pulses in a regular rhythm, alternating between closely spaced pulses (mmmmmmm..zi..zi..zi..zi..zi), more widely spaced pulses (mmmmmmm..zi....zi....zi....zi....zi), and back again (mmmmmmm-..zi..zi..zi..zi..zi) in repeating cycles.

Males of the most commonly studied *Drosophila* species normally produce pulses spaced thirty-five milliseconds apart and vary their songs between closely and widely spaced pulses in cycles of about sixty seconds. But several mutations in a single gene alter that rhythm in different ways. Called the *period* gene or *per*, it has three mutant forms: the allele *per*-long causes males that bear it to produce their songs on an eighty-second cycle, the allele *per*-short causes them to sing on a forty-second cycle, and the allele *per*-0 robs them of all rhythmicity.

Interestingly, mutations in the *per* gene also knock flies off their normal behavior in another way, by disturbing normal sleep/wake cycles. Fruit flies, like humans and many other animals, have an internal biological clock that sets the timing for twenty-four-hour patterns of daily rest and activity, even when flies are kept in continuous darkness. *Per*-long flies, however, have daily cycles of twenty-eight hours, *per*-short mutants have cycles of nineteen hours, and *per*-0 mutants are, as Kyriacou puts it, "insomniacs"—they have no regular activity cycles at all.

The laboratory *per* mutants are decidedly abnormal; *per*-0, in particular, is so behaviorally out of kilter that it would probably not survive in nature. But finding those mutant alleles enabled researchers to isolate and identify the "normal" version of a gene involved in running *Drosophila*'s biological clock. That accomplishment made it possible to analyze natural variants of the gene in wild populations across Europe, the Mediterranean, and northern Africa. The survey, by uncovering several alleles in wild populations, revealed that various forms of the gene are found in different proportions in different places. This raises the possibility that these natural variants serve some role in the animal's behavioral adaptation to varying environmental conditions, a hypothesis that is being tested.

Several other mutations in laboratory fruit fly strains have been shown to affect the animals' ability to learn. Of course, even under the best of circumstances, *Drosophila* can't handle calculus. But normal flies can learn that a specific odor signals the imminent arrival of an electrical shock. Flies with mutations named *dunce* or *rutabaga,* however, can't learn that association, and another mutant, *amnesiac,* can learn—but can't remember for very long.

Several of these fruit fly learning mutants are being reexamined to determine how—or even if—these genes interfere with the process of learning *sensu strictu.* Some mutations seem to affect either the flies' sensory systems or their motor systems rather than their associative, decision-making, or memory functions. Although such mutations are useful research tools (because they shed light on molecular mechanisms behind sensory and motor systems), they don't involve learning and memory directly. One could hardly say that a blind organism has a memory problem because it can't recall a visual stimulus that it cannot see! To really show that a gene influences learning, a researcher must show that its actions affect the neurons that process sensory information, compare it with data stored in memory, decide what to do about the situation, or enable the animal to retain its training.

Because a few known *Drosophila* learning mutants do seem to involve this third function, they are potential keys to questions about memory. The product of the *dunce* gene, for example, is an enzyme known to be involved with learning and memory in other animals. What's more, although the enzyme is found throughout the nervous system, the *dunce* gene itself seems to make its product most actively in parts of the fly brain called mushroom bodies, structures implicated in learning and memory.

Animals with more complex nervous systems are much more difficult subjects for this type of precise, "one-gene-one-effect" behavioral genetics work. Only a few known genes have well-defined effects in humans, and those are involved with severe neurological disorders rather than with more normal variations in brain function. One dramatic example is Lesch-Nyhan syndrome, a tragic, single-gene disorder. People afflicted with it are hampered by mental retardation and by a powerful urge for self-mutilation; no sooner do their teeth grow in than they begin to chew their tongues, lips, and fingers. They are also likely to bite their caretakers.

Another single-mutation affliction is phenylketonuria (PKU), which appears in individuals who carry two defective copies of a single, vital gene whose product breaks down phenylalanine, an amino acid found in many foods and certain artificial sweeteners. Because that amino acid is not broken down in their bodies, PKU patients can be plagued by abnormal brain development and mental retardation, especially during infancy and early childhood.

Lesch-Nyhan and PKU are pathological conditions, so investigating their molecular underpinnings, though medically valuable, has more to do with identifying singular genetic defects than it does with studying the effects of genes on ordinary, day-to-day behavior. Finding and working with genes like natural *per* mutants is extremely difficult in mammals, in part because many of our behaviors seem to be influenced by several interacting genes rather than just one or two. To tease apart those interactions, breeding experiments are essential; a team at Berkeley, nicknamed the Dog Genome Project, is now cross-breeding Border collies with Newfoundlands to isolate genetic influences behind those breeds' distinctive behaviors.

YET ANOTHER FACTOR complicates efforts to link genes and behaviors: interactions between genes and environment. Although genetic instructions *shape* certain traits, genes don't necessarily *determine* those traits irrevocably. Why? Because genetic instructions and environmental influences interact constantly during growth and development. Those interactions can profoundly affect the final characteristics of traits ranging from physical features to fine-tuning of the nervous system. Several simple examples make this point in different ways.

Most fishes have two eyes, a trait "controlled" by genes that govern development of the head and the nervous system and differentiation of the left and right sides of the body. But in the early 1900s, one experimenter raised fish embryos in seawater containing a high concentration of magnesium ions, with unexpected results. Some embryos, which carried the same genes for eye development as normal fishes, developed a single, cyclopean eye instead of the normal pair. What had happened? The change in magnesium concentration didn't cause any mutations in DNA. Instead, in a manner still not understood, a change in environmental conditions caused a dramatic change in *gene expression* during development. Genes which "determined"—irrevocably, one would have thought—that developing larvae grow two eyes, no longer acted the same way to produce the same trait.

An equally dramatic effect is found in certain human conditions. The defective genes that cause PKU, for example, once invariably led to severe mental retardation. Then researchers identified the problem as an inability to metabolize phenylalanine. Armed with that knowledge, physicians can now identify phenylketonurics early in infancy and place them on phenylalanine-free diets. This simple dietary change dramatically alters their genetic "destiny" by enabling them to mature into perfectly normal, healthy adults.

Notice what has happened here. These individuals carry precisely the same genetic flaws that would once have doomed them to retardation, but as long as they steer clear of phenylalanine, they have a much different prognosis. Now, PKU is a highly heritable condition. It is, in fact, 100 percent heritable; homozygotes for the defective gene cannot metabolize that amino acid. But the end results of that inability can vary enormously, depending on environmental conditions.

These widespread sorts of environmental effects on gene expression profoundly complicate attempts to evaluate what genes do. Suppose, for instance, that you are a geneticist trying to evaluate the influence of genes on height in a plant species. To start, you collect a sack full of seeds. If you sow those seeds under controlled conditions and expose each seedling to precisely the same environmental conditions—sun, water, nutrients, and so on—the resulting plants will exhibit some average height, with some being shorter and others being taller than average. You can then perform breeding and gene-mapping experiments to link various alleles to height, much as Mendel did with his tall and short peas. In this situation, you can probably find a strong correlation between particular alleles and plant height.

Now, however, you perform a different experiment: you pick two handfuls of seed from your sack at random and sow one handful in rich, fertile soil and the other in poor, sandy soil. Not surprisingly, the plants grown in rich soil grow taller than those grown in poor soil. Each of those populations, however, still contains some individuals that are taller and others that are shorter than their fellows. By conducting studies within each population, you can find the same strong correlations between specific alleles and plant height. But your analysis of genetic effects *within* those separate populations can't tell you anything about the relationship of genes to the difference in height *between* those populations. Why? Because that difference is profoundly influenced by environmental differences.

Talk of seeds, soil, and plant height may seem divorced from our topic, but it isn't, because gene-environment interactions are integral to the growth and function of the nervous system, and hence to the emergence of behavior. Consider first the 100 billion neurons in the human brain, each of which typically passes information to as many as 1,000 of its fellows, and receives input from as many as 10,000 more. There couldn't possibly be enough information in the human genome to direct the wiring of each and every connection between those cells. Instead, it seems that genes map out general "wiring diagrams" for the nervous system by producing trails of chemical markers. Typically, the growing tip of a young neuron "sniffs" its way along a trail of those marker molecules like a bloodhound searching for prey. Ultimately, it locates its genetically programmed attachment point, connects with other neurons in the vicinity, and begins to communicate with them.

But from that point on, the messages the neuron receives and passes along depend on activity in the developing brain, which in turn depends on the experiences of the animal. And critically, differences in those experiences—which result in different patterns of brain activity—determine which neurons remain alive and how their interconnections are finalized. This phenomenon was demonstrated through experiments with cats, in which the final wiring between eye and brain is not completed until after birth. It turns out that if kittens are raised in artificial environments containing nothing but horizontal lines, their brains fail to develop and maintain the neural connections that enable them to see vertical objects. Such animals often bump into chair legs because their nervous systems have been permanently altered, not by genetic anomalies, but by unusual environmental conditions.

This powerful effect is only part of environment's potential influence on brain and behavior. In many animals, neural circuitry retains some flexibility throughout life, making it possible for experience to modify existing patterns of behavior through learning—a process that often involves changes in behavior once labeled as genetically hard wired. Male fruit flies, though directed by genes in the mechanics of courtship serenades, can learn to stop courting females who don't respond. Human infants, for their part, begin suckling according to prewired directions.

But as nursing mothers can attest, babies refine their activities and become more efficient with practice. All chimpanzees may begin life with similar innate predispositions to particular styles of hunting, methods of gathering food, and the use of tools. But rain forest–dwelling chimps of the Ivory Coast use different tools and hunt differently than their counterparts in the drier environments of Tanzania. The more complex a trait, the more involved and obscure the effects on it of learning and other environmental influences can be.

Inbred strains of laboratory mice, for example, exhibit marked differences in aggressive behavior that can be measured by rigorous, carefully controlled tests. Numerous researchers, certain that these differences result from genetic variations among strains, are conducting elaborate experiments to map and identify alleles behind them. Those experiments involve cross-breeding strains with different levels of aggression and "backcrossing" resulting hybrids (breeding them back to one or the other of their parents) to identify genetic markers associated with aggressive behaviors.

Nonetheless, researchers who have dedicated their professional lives to documenting the influence of genes on behavior know that they must monitor their subjects' environment with precision that an outsider might consider positively obsessive. Why? Because the behavior of an individual mouse in a test can be affected by the time of day and season of the year at which testing is performed, the amount of light in the home cage and test arena, the size of the test arena, and the duration of the test. Mice transferred from cage to cage with forceps behave less aggressively than those carried around in small boxes. Males reared in the presence of an adult male have higher levels of offensive aggressive behavior than those reared by their mother alone. The number of littermates a mouse had, the number and type of animals it has been housed with, and the nature of its food and drink can also influence levels of aggression. In fact, environmental influences begin in the womb itself: male fetuses positioned in the uterus between two brothers score higher on certain aggressiveness tests than fetuses surrounded by sisters.

What's the message here? Can genes exert demonstrable influence on aggressive behavior in strains of mice? Substantial evidence indicates that they can. Does a particular combination of genes determine irrevocably the way an individual mouse will behave? Not on your life! "Genetic influence," "environmental influences," and "gene-environment interactions" are statistical terms determined by the study of groups of individuals. As mouse researchers know, observation of genetic effects on populations (groups of mice) are not easily translated into predictions of the way an individual mouse with a particular genetic makeup will behave in a given situation at a specific time. This difficulty is one reason why, after almost fifty years of work, mouse researchers still know little or nothing about specific genes involved in aggressive behavior, how those genes interact during development, or how they change as behavior evolves in various strains. And they are nowhere near isolating a specific "aggression gene."

• • •

LESSONS FROM THESE animal studies help explain why studies in human behavior genetics have been so equivocal and disappointing to date, despite a promising series of research reports issued several years ago. In 1987, one study linked a marker on chromosome 11 to the presence of manic-depressive disorder among individuals in an Amish farming community. In 1988, another study implicated part of chromosome 5 in schizophrenia. Both these teams hoped to intensify their searches to find the specific genes involved. Then, in 1990, Noble and Blum announced their putative "alcoholism gene."

The notion of genetic underpinnings for these ailments made—and still makes—sense to many researchers. Both schizophrenia and manic depression cluster in families, and identical twins are significantly more likely than fraternal twins to share diagnoses of schizophrenia. Alcoholics Anonymous and other proponents of the alcoholism-as-disease theory have insisted for years that the condition must be inherited. Moreover, none of these studies proposed irrevocable outcomes for people carrying certain genes. Rather, they suggested that certain genes create predisposition or vulnerability to particular mental states, but that one or more environmental triggers are necessary to cause disease. That notion fit well with emerging views of several ailments.

Most multiple personality cases, for example, are linked to severe emotional, physical, and/or sexual abuse, especially during early childhood. But the converse relation doesn't hold: the vast majority of abused children do *not* end up with multiple personality disorder, although they often have other serious psychological problems. So it is logical to hypothesize that while the condition is precipitated by abuse, it develops only in abused individuals who are genetically predisposed to react to abuse in specific ways.

But the euphoria over these apparent genetic linkages was short-lived. The manic-depression study was disproved and retracted within two years. The schizophrenia results were called into question more quickly than that. And the alcoholism study was disputed as soon as it was published. These retractions and disputations were duly noted, but not prominently featured, in the scientific press. They were not well covered at all by popular media; if newspapers and magazines published stories about the retractions, they were buried far from the front page.

What went wrong? Why were the results of these studies so tenuous? Part of the answer is simple: researchers can't breed people like mice, so they must rely on analyses of afflicted families. This approach restricts both the type of information that can be gathered and the number of subjects available. "I wouldn't even *think* of submitting a *Drosophila* paper with a sample size of fewer than a thousand at the very least," one behavioral geneticist told us. "In human behavioral studies it can take a decade to get a sample of a hundred, so we tend to be content with that."

But sample size can't be the only problem. After all, researchers tracking genes for genetic diseases such as cystic fibrosis, Alzheimer's, and Huntington's disease, though hampered by the same restrictions, are managing to find the genes they seek. One factor compounding the sample-size problem is the variable and idiosyncratic nature of both mental conditions and diagnoses. British and American psychiatrists, for example, often disagree on whether particular individuals should be diagnosed as schizophrenic. As Eric Lander of the Whitehead Institute has written, "There remains no universally accepted definition of schizophrenia, no central feature at the heart of the disorder (such as mood changes in manic depression) and no pathological underpinning (such as neurofibrillary tangles in Alzheimer's disease)."[11] In the case of alcoholism, Paul Billings adds, "I don't think that my definition of an alcoholic, a person from Utah's definition of an alcoholic, or a citizen of France's definition of an alcoholic are likely to be the same. I don't know how one can study the genetics of a phenomenon when we can't even agree on what the phenomenon is."

To make matters worse, some broadly defined disorders, including alcoholism and manic depression, contain several related conditions that grade into one another. The broadly labeled manic-depressive disorders are subdivided into bipolar I (mania), bipolar II (hypomania), schizo-affective disorder (manic-depressive types), and atypical psychosis with prominent affective features. Similarly, although family members may share a diagnosis of chronic alcoholism, they don't all necessarily react to alcohol in the same way.

Furthermore, both alcoholism and bipolar disorder manifest themselves over a wide age range. Thus, people with different conditions may receive similar diagnoses, and people with the same conditions may be diagnosed differently by different clinicians or at different times in their lives. This diagnostic ambiguity combines with many-faced disorders and restricted sample size to create formidable problems for gene hunters. The study of manic depression in the Amish, for example, was derailed when two individuals originally classified as normal were diagnosed with the disorder after the initial linkage study was completed. Those people just happened to be in the "wrong" part of the pedigree as far as the original results were concerned, and—partially because of the small size of the original sample—their conversion invalidated the results.

STILL ANOTHER DIFFICULTY in human behavioral genetics relates to the complexity of behavioral traits. Most genetic conditions discussed in earlier chapters are known as qualitative traits, because you either have them or you don't. Qualitative traits often come in two or three distinct "flavors," like the attributes Mendel studied in peas; depending on which genes his pea plants carried, their seeds were either smooth or wrinkled, and their flowers were red, white, or pink.

In humans, cystic fibrosis acts more or less like seed-coat texture in peas: homozygous normal individuals are unaffected, heterozygotes are carriers but show no symptoms, and homozygotes for the defective gene have the disease. Familial hypercholesterolemia, on the other hand, acts more like flower color: homozygous normal individuals have normal cholesterol levels, heterozygotes have cholesterol levels two to four times above normal, and homozygotes for the defective gene have levels six times normal or higher.

But many traits, including many behaviors, spread out over a broad spectrum rather than falling into two or three discrete categories. Humans, for example, can't be divided neatly into three intelligence groupings called geniuses, average folk, and imbeciles. Regardless of which "intelligence tests" are used, any large sample produces a broad distribution from very bright on one end to decidedly slow on the other. Traits such as these are known as quantitative traits. These are often influenced by several interacting genes, some of which have large effects and some of which have small effects. They are also influenced by several environmental factors, which can also vary in strength. (Remember the mouse aggression studies.)

The key word here is "interaction"; the number of potential interactions increases exponentially as the square of the number of factors involved. That's one reason quantitative traits can be devilishly difficult to study. To begin with, researchers rarely know in advance how many genes affect a given trait. That alone isn't an insurmountable obstacle, but it is possible that different genes affect the trait in different individuals or populations. It is also possible that the effects of specific genes vary from one individual or population to another because of the presence or absence of still other genes. Layered on top of that uncertainty are the effects of environmental variables.

When so many unknown interacting factors may be present, the results of their interactions may be largely unpredictable. If researchers know nothing in advance about any of the genes or environmental factors involved in a particular trait, it is eminently possible for them to develop hypotheses that can be "proved conclusively" by one small data set and categorically disproved by another. That flaw in experimental design, in one way or another, was the Achilles' heel of each genetic mental health study discussed above.[12]

Despite these obstacles, the genetic basis of quantitative traits can be studied through a technique called QTL (for *Quantitative Trait Loci*) analysis. One QTL study selected two species of tomato whose fruits differed from each other in several traits of interest to plant breeders: size, acidity, and a characteristic called soluble solids. By crossing those two species, backcrossing the resulting plants, and analyzing 273 progeny, the researchers located several genes that influence each of those traits. To be precise, they found six regions of the genome that contribute to fruit size, four to soluble solids, and five to fruit acidity. Similar studies are under way in several animals, including laboratory mice. A number of these studies, considered promising at first, have run into procedural and technical problems, and their preliminary results have been questioned.

Despite these difficulties, and despite withering criticism from researchers conversant with the possibilities and limitations of QTL analysis, several investigators are applying this technique to human behavioral genetics. Robert Plomin of Pennsylvania State University, for example, is searching for genes that influence intelligence. His project, which concentrates on 600 children from six to twelve years of age who range from "gifted" to mildly retarded, is being funded to the tune of $600,000 by the National Institute on Child Health and Human Development. In addition to performing a battery of cognitive tests on the young subjects, the research team will establish permanent laboratory cell lines from blood samples, so that DNA will be available for analysis at any time. Plomin and his colleagues have convinced themselves—and obviously, the scientific review panels of major granting agencies—that their approach circumvents, in a scientifically valid way, limitations that working with human subjects places on QTL analysis.

But investigators familiar with behavior genetics and with the intricacy of QTL vehemently denounce this kind of work and insist that these studies are fatally flawed. First, they point out that crossing and backcrossing are out of the question and that the genetic backgrounds of subject populations are not sufficiently well understood to permit rigorous mathematical analyses. They further argue that those limitations, formidable in themselves, become insurmountable when combined with the "fuzziness" and complexity of the behavioral "traits" under investigation. One eminent behavior genetics researcher—himself involved in analysis of complex traits—echoed these concerns in a private communication about the uncertainty of this work. "Most investigators studying the genetics of multifactorial traits," he wrote, "are not skilled at statistical analysis of genetic data . . . and are accepting linkage at confidence levels well below appropriate levels. This will have the unfortunate effect of littering the genetic map with loci that cannot be confirmed, because they were placed on the map without enough evidence."

Then there are the famous (or notorious, depending on your point of view) twin studies. These investigations focus on *monozygotic* (one-egg) twins, rare individuals who are produced from a single fertilized egg and begin life with essentially identical genes. Many such studies, in an effort to distinguish genetic influences from environmental contributions to human behavior, compare monozygotic twins reared *t*ogether (MZTs) with monozygotic twins separated at birth and reared *a*part (MZAs). The best of them also include comparisons with either fraternal (*dizygotic,* or two-egg) twins, who are no more genetically similar to one another than nontwin siblings or to unrelated children adopted into the same family. By their nature, twin studies must examine statistical effects of genes in populations rather than gene action in individuals.

Twin studies often differ from one another in the specific psychological tests employed to measure personality and intelligence. Their results vary in terms of the precise percentages they assign to genetic and environmental influences on different traits. But they generally concur in several conclusions: that MZTs are surprisingly similar to one another in behavior and personality measures ranging from IQ to shyness, that MZAs are extremely similar to MZTs in most of those measures, and that adopted children reared in the same household bear no more resemblance to one another than people selected from the population at random.

In the eyes of the researchers who conduct these studies (eyes, which critics argue, are intrinsically biased), the implications are clear; somewhere between 30 and 70 percent of the variation in behaviors of identical twins is attributable to genetic variation rather than to differences in environment. In an article in *Science,* psychologist Thomas Bouchard has asserted, "For almost every behavioral trait so far investigated, from reaction time to religiosity, an important fraction of the variation among people turns out to be associated with genetic variation. This fact need no longer be subject to debate; rather, it is time instead to consider its implications."[13]

Robert Plomin concurs. "It must seem very far out to you, but of various things we can talk about, like heart disorders and other complex medical disorders, there's better evidence of genetic influence on cognitive ability. It's not enough to say that genes are important. Adoption data and twin data all say that cognitive ability and IQ have very substantial genetic influence, more than any complex trait other than height."

But many prominent geneticists and experts in animal behavior, though acknowledging general genetic influences on behavior, get positively apoplectic about such categorical statements regarding genetic control or determination of complex behavioral traits by specific genes. These critics roundly dismiss such claims as unverifiable, even ridiculous, assertions, emphasize that they are based on fundamental misunderstandings of population genetics, insist that they involve egregious errors in extrapolating population statistics to individuals, and assert that they disregard pervasive environmental influences.

Among the fatal flaws of human twin studies are their necessary lack of environmental controls and interventions—linchpins of the experimental method. Researchers can neither selectively breed human subjects nor exert control over their environments. Critically, similarities in environments shared by twins reared apart are often glossed over by twin study supporters. Mark Newman and Gerald Levey, our "identical firemen," both grew up in middle-class, Jewish homes in the New York metropolitan area—Levey was raised in Brooklyn, New York, while Newman grew up in Paramus, New Jersey. As Bouchard himself admits, most twins involved in most studies were brought up under the influence of predominantly white, middle-class families in modern Western society. Among other factors, adoption agencies tend to select fairly similar sorts of families to receive their children.

The unavoidable lack of control over rearing environment places serious limitations on the inferences that can legitimately be drawn from their data. Take a hypothetical claim, based on twins like Newman and Levey, that a gene or set of genes "produces firemen"—that genetic influence takes precedence over environmental ones in shaping individuals' lives regardless of their life histories. "Take one of those kids and put him in a *really* different environment," suggested one behavior geneticist, "like in a family of bushmen in Africa, or in a farming village in mainland China, and *then* come back twenty years later and see if you find two firemen who dress the same!"

If you recall the number and variety of variables that can affect the behavior of mice in aggression tests, you can understand why Cold Spring Harbor Laboratory's Tim Tully throws his hands in the air at the prospect of behavior genetics studies in humans. "When you try to partition variability of a human population into nature and nurture," Tully points out, "the loss of experimental control introduces severe complications to interpreting the data, complications that people who still try to address human behavior genetics issues this way constantly choose to ignore or oversimplify."

But the loss of experimental control is only one difficulty involved in comparing heritability of behaviors across populations and cultures. If you recall the examples of fish eyes, plant height, and PKU, you can appreciate a statement made by Harvard University population geneticist Richard Lewontin more than a decade ago. "There is no such thing as *the* heritability of IQ," Lewontin has emphasized repeatedly, "since heritability of a trait is different in different populations at different times."

Let us assume for the sake of argument that twin studies are free from internal methodological flaws. *If* those studies are internally sound, the striking similarities in IQ and other behavioral measures between MZTs and MZAs *would* indicate a strong genetic influence on those traits *for those populations of twins*. They would also suggest a concomitantly weak influence of childhood environment on those populations. But that information would be directly relevant to the broader population if, and only if, the environments in which the twins developed were representative of the full range of human environments around the globe, and if the twins were themselves genetically representative of the population at large. Neither of these conditions seems to be true.

ALL THESE PROBLEMS and uncertainties notwithstanding, arguments around them would be little more than academic discussions if claims for genetic determinism concerned only such "traits" as preferences for Budweiser, John Wayne movies, and double-breasted shirts. But these studies make powerful predictions about the genetic control of such vaguely defined characteristics as intelligence, and that's where sloppy scientific thinking can have grave social consequences.

Patrick Buchanan, for example, as a White House aide in the 1970s, responded to reports on race and IQ with a memo to President Richard Nixon. Referring to an article by psychologist Richard Herrnstein in the *Atlantic Monthly,*[14] Buchanan wrote: "Basically, it demonstrates that heredity, rather than environment, determines intelligence—and that the more we proceed to provide everyone with a 'good environment,' surely the more heredity will become the dominant factor—in their intelligence, and thus in their success and social standing. It is almost the iron law of intelligence that is being propounded here—based on heredity."[15]

Buchanan then used his interpretation of the article to make a suggestion with which many behavior researchers would disagree—that efforts to improve the standing of blacks and the poor through integration, compensatory education, and providing an "equal chance at the starting line" might be doomed by the possible genetic inferiority of certain races and classes. When asked about the memo during the 1992 presidential campaign, Buchanan dismissed as "nonsense" the notion that he believed whites were genetically superior to blacks.[16]

The sort of reasoning Buchanan's argument exposes is flawed for several reasons. First and foremost, the notion that IQ tests somehow measure and describe with a single number the elusive and multifaceted quality we call intelligence is hotly contested. Yet intelligence researchers tend to accept its validity without qualification.

Second, IQ scores have been shown to be highly heritable only within pairs of identical twins, people who begin life not only with nearly identical genes, but with genomes organized more similarly than those of any other humans on earth. The overall workings of twin genomes, therefore, are much more similar than the genetic interactions in the genomes of any other pair of individuals. But even in that restricted and unrepresentative population, estimates of IQ heritability range from a high of 70 percent to a low of around 40 percent. Notice that those percentages still consign between 30 percent and 60 percent of the observed variation to environmental causes.

Third, the claim of IQ test designers that their instruments are culture-free—free of bias favoring individuals from certain classes or sociological background over others—is very much open to question. The notion that IQ tests administered to whites and African-Americans measure nothing besides differences in innate intellectual ability is even more contentious. Even if the contents of the tests themselves could somehow be considered culture-free, another important question remains: Can anyone believe that the majority of whites and the majority of African-Americans enter such tests with similar life experiences? The lack of environmental control in human studies rears its head once again.

Finally, the notion that differences in IQ between heterogeneous black and white populations are caused by genetic differences between those populations remains unsupported by scientific evidence. No genes have been found whose presence or absence can be used to test the hypothesis, and no data exist showing heritability of IQ scores between those two populations. Remember the case of seeds sown under different conditions, different environments, affecting the expression of the same genes, can produce widely divergent results.

These daunting problems haven't prevented continuations and extensions of these studies, partly because so many molecular biologists believe in their reductionist vision. Predicts one eminent cancer researcher, "We are going to find, ten or twenty years from now—I can't predict how and when—that there are certain genes that are determinants of intelligence, however you want to define that. People have shrunk back from describing or quantifying intelligence because our quantifiers now are so poor. But that doesn't mean there aren't people who are born smarter than others. It only means that the measuring tools are defective."

In the meantime, the search for genes that influence human behavior has been expanded to traits including conformity, aggressiveness, and ambition. More to the point, the same limited data have been boldly extrapolated to predict the extent of genetic influence on behavior in the general population. Estimates published by the Minnesota group range from a high of 61 percent for extroversion to a low of 33 percent for intimacy. Once again, a combination of irresponsible speculation, careless reporting, and the swing of scientific opinion toward genetic determinism have created a slippery slope that threatens to fuse the clear-cut genetics of sickle-cell anemia with inherent tendencies toward criminality, alcoholism, domestic and social violence, and homelessness.

• • •

ARE THERE ANY PATHS through these technical, analytical, and social obstacles to a better understanding of the relationships between genes, environment, and the human brain? What have we to look forward to as researchers from different fields attempt to meet on the molecular level? Fortunately, the evolutionary history of our brain is very much within the brain itself; similarities in structure and function unite both nerve cells and nerve networks across the animal kingdom. "Therefore," Tim Tully points out, "if we can define a learning paradigm in flies, and identify genes that are involved in the production and appearance of learning in flies, a very large number—I would say a majority—of those genes are going to underlie aspects of learning and behavior in humans."

Tully and a growing group of molecular neurobiologists are concentrating on identifying proteins that underlie learning, pinpointing genes that produce and control those proteins, and examining how genes and proteins function on a cellular level. Their next step will be to discover how those proteins affect the actions of neural networks, and ultimately how those networks affect behavior.

That approach is slow and painstaking, but fruitful, as evidenced by work on several animals. Researchers working with one marine snail, for example, are learning how genes control egg laying, a fixed behavior pattern somewhat similar to the egg-retrieving behavior of the greylag goose. Recently they identified a family of genes which code for several compounds that prompt the snail's nervous system to perform the behavior. Two teams working with other marine snails have, in parallel—though not always in agreement—been zeroing in on genes and gene products associated with the physical and chemical changes in nerve cells that underlie learning and memory.

Tully, meanwhile, is collaborating with colleagues to combine molecular biology with the sophisticated genetic work for which fruit flies are ideal subjects. One team member cloned a gene that looked similar to a gene in humans. The human gene produces amyloid protein, a compound closely associated with the memory loss of Alzheimer's disease.

Using tricks of genetic manipulation in flies, Tully and his colleagues produced flies carrying all the genes in their normal genome, except for that particular gene. Those flies survived well in the laboratory, were fertile, and appeared perfectly normal. But when tested in certain behavioral experiments, they showed significant decreases in performance. Finally, when the team inserted a copy of the *human* gene for the related protein into those flies, their behavioral performance recovered significantly.

These experiments have been unable to demonstrate conclusively that this gene affects learning and memory, because it also affects certain aspects of the flies' escape response that don't involve those functions. But the fact that a protein known to affect the central nervous system in humans has been found to have related effects on flies is a potential godsend. Why? Because experiments on these animals, which can be selectively bred and subjected to a wide range of controlled treatments, can help describe in intimate detail the functioning of genes that are important, but extremely difficult to study, in humans.

Experiments of this type have other advantages as well. Tully argues, "When you apply molecular biology to analyze the mechanism of behavior, not the population variability of behavior, which is what the nature/nurture thing is about, you actually gain some insight on how behavior is expressed in *individual* animals rather than in populations."

WHAT, IN THE MEANTIME, are we to do with studies of genes and human behavior that work "from thc top down" by trying to identify genetic controls from human subjects by use of the techniques discussed earlier? Clearly, one must be exceptionally wary of snap judgments about their significance—far more wary than either the general scientific press or popular media would have us believe.

Take the hotly disputed alcoholism studies. On the surface, they seem quite solid. The gene that Noble first fingered as a major factor in alcoholism makes a nerve cell receptor that enables the brain to respond to a neurotransmitter substance known as dopamine. Dopamine, in turn, seems to be important in the neural mechanisms involved with reward, a brain response that reinforces behaviors which an individual finds pleasurable. Prior to Noble's study, there was already evidence that alcohol affects the dopamine centers of the brain, and that dopamine is involved somehow in "rewarding" the effects of alcohol consumption, although several other neurotransmitters were known to be involved as well.

With this information as background, Noble and his colleagues determined that there were two forms of the dopamine receptor gene, alleles named A-1 and A-2. When they compared the brains of people diagnosed as alcoholics with those of others not so diagnosed, they found that the alcoholic group had what they described as a significantly higher occurrence of the A-1 allele. They also determined that these individuals had fewer dopamine receptors in certain important parts of their brain. "And we figured from that basis," Noble explains, "that these people had a deficiency state in their brains. They didn't have enough of these receptors which, when stimulated, give people pleasure." Lacking normal responses to pleasurable stimuli, the hypothesis continues, these "receptor deficient" individuals need more stimulation than normal people, a need that alcohol fills by stimulating the release of excess dopamine.

Since the publication of this original hypothesis, which was too often reported as fact, subsequent investigations have shown that although the A-1 allele occurs more frequently in certain types of alcoholics, it also occurs in people who don't suffer from the disorder, and doesn't necessarily occur in all members of a single family afflicted with alcoholism. Consequently, proponents of the significance of this gene have broadened their hypothesis to include influence on several other habits.

"I don't personally feel that this is an alcoholism gene per se," Noble later explained, showing a marked change in emphasis from his earlier statements. "I think this gene is involved in a number of behaviors. We believe that this gene manifests itself in some people with alcoholism, in others with cocaine addiction, and perhaps in others with disorders like smoking. Nicotine is a great releaser of dopamine; it's like alcohol. So it gives you a charge. It gives you a high. And food is the same way, perhaps. High amounts of carbohydrates, like chocolate, also release dopamine in the brain. We believe that this gene might subserve a lot of behaviors which we call bad habits. So we are beginning to look at the prevalence of this gene in other disorders."

In addition, Noble now hypothesizes that the gene is involved in a predisposition toward a personality trait loosely defined as "risk taking." Alcoholics and their children, Noble explains, "tend to be loners. But they also, interestingly enough, have compulsive behaviors. When they do something, they stick to it. For example, they typically choose activities which we call risk taking. They put themselves in a situation of great excitement just to stimulate themselves—perhaps to release dopamine and stimulate their receptors. So they take a lot of chances, a lot of risks."

Implicit in this line of reasoning is the judgment that risk taking, however it is defined, is somehow undesirable or, at the very least, frivolous.

That logic strikes a responsive chord in Kim Collins, who sees compulsive and risk-taking behavior in both her husband Gary and their son. "They both want excitement, stimulation all the time. They like to push things to the limit. Gary does everything excessively. You know, he doesn't drink alcohol now, but he drinks a lot of coffee, he smokes a lot of cigarettes. I see that sort of excessive behavior in Ian, too. It's very similar, and it scares me to think that if Ian does experiment with alcohol, it would grab hold of him like it did Gary."

At first glimpse, it's easy to share Kim Collins's concern. Ian is currently enamored of bungee jumping, an activity that not everyone finds sensible. But who is to define where "risk taking" begins? Several researchers, though not Noble, hypothesized that our twin firemen share a gene for risk taking that influenced their fondness for firefighting. Being a volunteer fireman in one's spare time does involve more risk than gardening. But is that pathological? And Gary Gross, who works with delinquent teenagers at a California state youth center, sees a connection between his need for stimulation and his chosen career. "I think one of the reasons I work where I do," he reflects, "is because of the stimulation I get from it. Working around troubled kids is exciting. It gives me the level of excitement that I want. They are unpredictable. They are a challenge. As I meet their needs as a teacher, they are meeting my needs for self-fulfillment."

Thus, the original idea of a single "alcoholism gene" has broadened and expanded to cover a wide range of behaviors, several of which are not exclusively connected with drinking, and others which may or may not be related to pathological behaviors. Even if we agree for the sake of argument that the dopamine receptor gene has something to do with stimulation-seeking, we must recognize that a host of environmental influences shape individuals' idiosyncratic responses to their "inherent need for excitement." We must also admit that there may be scores of other genes that moderate this single gene's effects, directing behavior in different directions at different times in different individuals. But if that is the case, the gene, or group of genes, in question might more accurately be said to heighten the effects of substance abuse or compulsive behavior, rather than causing them in the first place. "The environment is a tremendously powerful agent in producing alcoholism," Noble admitted in an interview with *Science News*. "But genes are easier to study."[17]

Despite all these changes in the notion of what the A-1 allele does, and despite the fact that no evidence yet suggests that any other single gene is responsible for "making" an alcoholic, researchers, members of the press, and people like the Gross family still talk about a genetic test for alcoholism. "If they came up with a test for alcoholism," Kim Collins states with little hesitation, "I think I would have my kids tested for that. At the earliest possible time, so that you could educate them, from the very beginning, that they could never drink."

"It is entirely possible to test for this gene," Noble asserts. "All we need is a small blood sample. We already do that routinely in the laboratories, so it is feasible. We are not yet testing patients. It's all research at this point. But I think soon one can move ahead to do this on a countrywide basis."

Exactly what would such a testing program identify? Alcoholics? Potential alcoholics? People with a susceptibility to addictive behavior? Individuals who tend to do everything "to excess"? Or otherwise perfectly normal men and women who choose stimulating careers and hobbies? Who would have access to this information, and what use could they make of it?

As we've discussed in earlier chapters, it is difficult enough to direct intervention and avoid discrimination around genetic testing when the science is solid enough to make certain a test is both accurate and specific. In cases where what is being tested is so nebulous and uncertain, do the potential benefits of testing outweigh the risks? One must question whether widespread testing should be encouraged, or even allowed, under these circumstances.

We thus return to the fundamental issues raised by virtually every other extension of molecular biology into the social arena, but this time with extra force, because the notion of genes controlling intellect, personality, and ability comes uncomfortably close to the characteristics by which we define our intrinsic worth as human beings. Here, questions of evaluating similarity and difference and concerns about productive and destructive social action arise with a vengeance.

Robert Weinberg of the Whitehead Institute, speaking with deep concern, articulates a conviction held by many molecular biologists who are not themselves experts in behavior genetics. "One of our ideological preconceptions in this country," he observes, "is that all men are created equal. That cannot, from the point of view of biologists, be the case. Not only will we find that people are genetically very different from one another, as we obviously must intuit by now, we will find that certain groups, ethnic groups, will be more or less well endowed with certain kinds of genes. It has to be the case from the point of view of the biologist. From the point of view of a democratic ideologist, I am horrified by that notion. But that is the future, inescapably, and we have to deal with it at present."

The discomfort here, as eloquently explained by François Jacob in *The Possible and the Actual,* comes from the impossible attempt to equate "two quite distinct notions: identity and equality. The former refers to the physical and mental properties of individuals, the latter to their social and legal rights. One is a matter for biology and education, the other for ethics and politics. Equality is not a biological concept." Yet, as Jacob adds, we find ourselves caught in a desperate struggle "either to base equality on identity or to justify inequality by diversity."[18]

You have seen repeatedly throughout this book that no two humans, except for identical twins, carry the same genes. (And just try suggesting to such twins that their DNA makes them identical people.) Every human on this planet is unique, and we should be thankful for that variety. Philosophically, to deny that people are different from one another is to deny an essential element of our humanity. Biologically, diversity is the key to evolutionary change and adaptation; without it, the biosphere as we know it would not exist. Accepting that diversity leaves us with two great challenges: one that must be faced by society at large, and the other that must be handled at the interface between science and society.

In the social arena we must work ceaselessly to ensure equality in the face of diversity. Even if, as some researchers believe, children's talents and abilities are extensively molded by their genes, that doesn't mean that parenting is inconsequential or that kids are mere marionettes manipulated by the tangled strings of the double helix. We should therefore do our best to ensure that all children have equal opportunities to become the best they can be, whatever their inclinations.

Meanwhile, molecular biologists must ensure that preliminary "findings" and unverified speculations neither generate ill-conceived social programs nor derail well-intentioned ones. For the most fundamental problem with currently popular investigations of genes and human behavior is that they can estimate only the proportion of the variance in sample populations that is caused by genetic factors. By nature of their design, these studies cannot evaluate, and therefore cannot pass judgment on, the power of environmental intervention to influence personality, intelligence, or specific skills. Most important, since these investigations do nothing useful to pinpoint the nature of gene action, they do not suggest further experiments that investigate the feasibility of specific environmental interventions.

The point here is not to deny that heredity influences behavior. "We who study behavior genetics are definitely hereditarians," Tim Tully says matter-of-factly. "We definitely believe that genes produce products that dictate the development and ultimate expression of adult form, of adult brain, and of adult behavior. But as behavior geneticists, we recognize that there is lots of environmental interaction with these gene products. The main thrust of our work should not be merely to show that genes are involved with behavior. That's trivial. Instead, we should try to understand how a gene product acts to produce a behavior. Then we might be able to intervene to correct or fine-tune behavior. That is not trivial."

ULTIMATELY, THE IMPACT of molecular biology on the study of behavior will be determined by our ability to face the prospect that the reductionist vision may become reality, that someday we may be able to determine, sequence, and develop tests for genes that powerfully influence the development of intelligence and other complex traits. What will we do if we can consult our genes, treating base sequences in DNA like molecular oracles? "I can't, science can't, tell you that," Robert Weinberg intones gravely. "But science in the form of one scientist, if not the whole field of science, can say that we as a society are not yet prepared for this deluge of information."

Facing such unpreparedness, Tim Tully knows his role. "It is not just that we have to be careful," he muses. "Our generation is going to fight the battle of genetic determinism again. It's inevitable, because of the tendency to construct pseudoscientific arguments supporting determinism and because of the current ignorance of molecular biologists about the sociological history of the problem. There will be many cases in the next twenty to fifty years where molecular biologists will jump to conclusions that smack of genetic determinism. And with the advent of molecular biology as a multibillion-dollar industry, molecular genetic determinism will be a powerful voice. It will influence social policy in a big way."

But the entire mind-set that treats genes as oracles is fundamentally mistaken. For the "prophecies" we may find in our genes are fundamentally different, *must* be different, from the prophecies of soothsayers and fortunetellers to whom humans have turned in the past. DNA in and of itself, to paraphrase Harvard's Richard Lewontin, is intelligence without substance, a script without actors, a factory with neither workers nor raw materials. It comes alive, fulfills its role as blueprint, and perpetuates itself only in the context of living cells—collections of materials, energy, sensors, and effectors. And those cells grow, develop, and change, not in a vacuum, but in an environment with profound effects on the results of genetic instructions.

Therefore, the information carried in our genes is not a prophecy but a blueprint. To the extent that reductionist science empowers our understanding, it can transform DNA from an inscrutable power that imposes itself upon us into a molecular intelligence we can learn to understand and employ to mutual benefit. What is needed, Tully argues, is a strategy to turn determinism on its head. "Molecular studies of behavior," he proposes, "will ultimately show that environmental intervention *will* work. It then becomes an issue altogether different from determinism—'There's something wrong with him, cancel his insurance, sterilize him, discriminate against him.' The issue is whether or not society will pay for interventionist strategies that solve particular problems. That may be a matter of national policy, or it may be a question of free-market economics. But it is not genetic determinism."

Our task is clear. We must choose and plan our questions carefully, ensuring that whatever information we gather will be put to use in the best interests of humankind. Some answers to current questions may ultimately tell us more about who and what we really are than some of us want to know. But what we choose to do with that information will tell us even more.

Notes

CHAPTER I

1. Horace Freeland Judson, *The Eighth Day of Creation* (New York: Simon and Schuster, 1979), p. 28.

2. Thomas F. Lee, *The Human Genome Project* (New York: Plenum Publishing Co., 1991), p. 41.

3. Judson, *The Eighth Day of Creation,* p. 30.

4. James D. Watson, *The Double Helix* (New York: Atheneum Publishers, 1978), p. 107.

5. It has been alleged that Rosalind Franklin did not receive due credit for her work. If you read *The Double Helix,* you should also read Ann Sayre's *Rosalind Franklin and DNA*. Franklin could not have shared the Nobel Prize for a technical reason: she died of cancer in 1957, several years before the prize was awarded. Nobel regulations state that the award may not be granted posthumously, so Watson and Crick shared the honor with Franklin's associate, Maurice Wilkins.

6. We recognize that this is not universally true. Some genes contain only fragments of the sorts of sentences we imagine here. Others contain several "run-on" sentences whose messages are separated at a later stage in information processing. Nonetheless, the analogy drawn is useful, and we ask knowledgeable skeptics to humor us by allowing its use.

7. Biochemists even have a name for these functional groups; they call them *domains*.

8. Conventionally, the genetic code is presented in the form of triplets located on messenger RNA. Because it suits our purpose here to discuss the triplet coding system at the level of DNA—a step earlier in the process—we have translated RNA codes for alanine (CGA) and leucine (CAG) "backward" one step.

9. *Atlas of Protein Sequence and Structure,* vol. 5 (Silver Springs: National Biomedical Research Foundation, 1972).

CHAPTER 2

1. James B. Herrick, "Peculiar elongated and sickle-shaped red blood corpuscles in a case of severe anemia," *Archives of Internal Medicine,* vol. 6, pp. 517–521.

2. Charles Darwin, *The Origin of Species* (New York: Penguin Books, Mentor Edition), p. 48.

3. Ibid., chapter four.

4. In a fascinating aside, Nagel describes the problem that faces African adults who travel abroad. It appears that acquired immunity against malaria, unlike immunity to smallpox or polio, is a transient, rather than permanent phenomenon; the immune system needs constant "reminders" to maintain its defense against these parasites. For that reason, Africans who leave malarial areas for as little as a year often become ill with malaria on returning home. These infections are not usually life threatening, but they can be serious.

5. Charles Darwin, *The Voyage of the Beagle* (London: Penguin Books, 1989), p. 367.

6. Charles Darwin, *Darwin's Journal of Researches into the Geology and Natural History of the Various Countries Visited by the H. M. S. Beagle under the Command of Captain Fitzroy from 1832–36,* 2nd edition (London: Murray, 1845). Contrary to popular impression, Darwin had no epiphany while stalking finches and tortoises in the Galápagos; animals there didn't hop into position to spell "evolution" on the

rocks. Darwin did recognize that many island species were found nowhere else, and he described the place as "a little world within itself" that was "very remarkable" in its flora and fauna. But his letters home from that part of the journey, rather than being illuminated with evolutionary insight, were filled with accounts of seasickness and homesickness. The paragraph quoted here wasn't published until 1845, a decade after his visit, and *Origin,* in which he elaborated on these ideas, wasn't published until 1859.

7. This study, which is methodologically similar to studies of mitochondrial DNA in animals, concentrated on the DNA carried in the *chloroplasts* of these plants' cells. Chloroplasts, like *mitochondria,* carry their own DNA, which does not participate in gene shuffling during sexual reproduction.

CHAPTER 3

1. In addition to testosterone, which directly influences the specific tissue transformations mentioned in the text, the testes produce a second hormone, Muellerian regression hormone, that influences which of two sets of reproductive ducts develops in the embryo. Later, in addition to the listed effects of testosterone at puberty, that hormone has significant effects on the central nervous system. In most other mammals, the intensity of male sex drive depends directly and exclusively on testosterone levels. In humans, however, many psychological factors are also important.

2. P. Koopman, J. Gubbay, N. Vivian, P. Goodfellow, R. Lovell-Badge, *Nature,* vol. 351, pp. 117–121, 1991.

3. J. Charfas, "Sex and the Single Gene," *Science,* vol. 252, p. 782, 1991.

4. Although the fear of such masquerades was that differential muscular response would give men unfair advantage, the only fully documented instance of such cheating in the Olympics doesn't corroborate that notion. The case involved one Hermann Ratjen, who claimed to have been forced by the Nazis to impersonate a woman in the high jump competition in the 1936 Olympics. Amusingly, Ratjen lost the event—to no fewer than three female competitors.

5. Their thesis, often considered the central lesson of molecular biology, was that gene expression is controlled by regulating transcription of genetic information from DNA to messenger RNA. This style of controlling gene expression is found in most organisms, although there are other ways to control protein production. Single-celled organisms known as trypanosomes, for example, make significant changes in messenger RNA *after* it has been assembled, a process that plays a role in controlling protein synthesis. It is not clear whether such processing exists in multicellular organisms. Some researchers dismiss that possibility out of hand, while others insist that RNA editing, as it is called, will be found in the human genome as well. In any case, there are no fewer than six places between transcription and protein synthesis at which regulation has been shown to occur.

6. Note that homeoboxes are not the only proteins that control gene expression. Another class of DNA-binding proteins has what are called zinc fingers, tendrils which, with the help of zinc atoms, curl around the double helix. DNA sequences that code for zinc fingers have been found in control genes of both fruit flies and humans. What's more, you have already met a zinc-finger protein in mammals—SRY! Still other classes of DNA-binding proteins include leucine zippers, helix-loop-helix, and helix-turn-helix. Thus, although homeobox-containing genes usually do seem to perform regulatory functions, all regulatory genes do not necessarily contain the homeobox sequence.

7. What looks like perfect external symmetry is broken by several minor inconsistencies, such as the fact that men's left testicle hangs a bit lower than the right one. In addition, quite a number of individuals have one leg or arm that is slightly longer than the other, as orthopedists and tailors will attest.

CHAPTER 4

1. It is not known whether or not the processes of wound healing and related cellular behaviors in cancer are exactly the same at the molecular level. The processes appear quite similar and probably overlap to some degree. In addition, the phenomenon of wound healing and subsequent restoration of normal cell behavior is a

reminder of the intricate series of controls and intercellular communications occurring in our bodies at all times.

2. The discovery of cancer-causing genes involves life forms ranging from viruses to humans. The first cancer-causing genes turned up in 1910, during studies of viruses that caused cancer in chickens, but it wasn't until the 1970s that researchers were able to manipulate the genetic information in viruses. Then a shocking truth came to light: the gene that caused cancer was not a part of the viral genome! Rather, it was a stretch of DNA that the viruses had picked up from a host cell and altered crucially. When viruses carrying copies of this mutated, kidnapped gene infected other cells, those cells became cancerous. The scientists who unraveled this story, Michael Bishop and Harold E. Vamus, were rewarded in 1989 with a Nobel Prize for physiology or medicine.

CHAPTER 5

1. We are talking here about response by European societies as a whole during the Middle Ages. To be fair, there were nearly always certain Christian fraternities who took it upon themselves to minister to the suffering.

2. The situation is actually much more complicated than this classification implies. There are many types and subtypes of T-cells, categorized according to the receptor molecules they carry on their surfaces. Several of these types may have more than one function. Note that many discussions of the immune system dating from only a year or two in the past refer to a third major functional class called suppressor T-cells, said to curtail the immune response at the end of an infection and to modulate its activity at other times. While it is clear that certain cells do perform these functions, it is now generally (and temporarily?) agreed that there is not a single class of T-cells that act exclusively as suppressors.

3. It turns out that "punctuation marks" in genetic sentences allow molecular "words" to rearrange themselves. Normally, that rear-

rangement happens only occasionally, and only during sexual reproduction, as discussed in chapter 2. In the case of antibody genes, however, the rearrangements take place as B-cells mature and differentiate within the body.

4. In reality, as this clone grows it continues the remarkable rearrangement of its antibody genes that produced the first match with a protein. The stretch of the gene that controls the "sticky" antibody region begins changing a million times faster than normal, as pieces jump around and switch places. As a result, each daughter cell carries an antibody-making gene slightly different from the version carried by its parent. Thus, when all the slightly different daughter cells make their antibodies and receptors, each creates a slightly different version of the molecule. Just by chance, some cells create antibodies that attach more efficiently to the particular viral protein that stimulated the original parent cell of the clone. Others produce antibodies that attach less strongly. Over time, a sort of evolutionary process takes place in this group of cells. Ultimately, those daughter cells whose antibodies match the viral protein most closely "win" and end up dominating the assemblage.

5. Similar population pressure and comparable intensive farming techniques exist over much of India and other parts of Southeast Asia as well. Outside China, however, pigs are often not part of the mixture. Religion plays a significant role in that phenomenon: in Islam, as in Judaism, pigs are considered unclean and unfit for human consumption. For various reasons, pork is also generally absent from the Hindu diet, and many Buddhists don't eat any meat at all. Thus, swine are not often raised in large numbers in many other densely populated areas of Asia.

6. They are, in fact, the genes that produce T-cell receptors in the same immunoglobulin superfamily as the genes that make antibodies and B-cell receptors.

7. M. H. Merson, "Slowing the Spread of HIV: Agenda for the 1990s," *Science*, vol. 260, pp. 1266–68, 1993.

CHAPTER 6

1. There are, of course, exceptions. A very few closely related plant species interbreed with one another in nature, and a larger number of related plant and animal species can be crossed artificially under conditions of domestication. "Tiglons" and "ligers," the progeny of matings between lions and tigers, for example, have been produced in zoos, but have never been found in nature. (And remember the cross between silverswords and tarweeds described in Chapter 3.) In many cases, however, the resulting interspecific hybrids are sterile; generically, they are described by a name borrowed from the most common example: mules. Fertile interspecific hybrids are few and far between.

2. This fairly crude method has disadvantages. Genes injected into eggs in this way are incorporated into egg DNA at random; they can end up inserting nearly anywhere in the genome. Sometimes injected genes end up in favorable spots on chromosomes and sometimes they don't. As mentioned briefly in Chapter 4, it seems as though a gene's location on a chromosome can have a significant effect on its expression, although the reasons for such effects are not at all clear.

3. Russ Hoyle, "Rifkin Resurgent," *Biotechnology*, vol. 10, November, 1992, pp. 1406–1407.

4. Ibid.

5. An article in *Sierra Magazine* compared Bromoxynil's toxicity to that of the herbicide Ordram. Ordram, which *Sierra* says has been responsible for massive fish kills, is toxic to fish at concentrations of 1.3 parts per *million*.

CHAPTER 7

1. The parallel between cystic fibrosis and sickle-cell anemia may actually go further. It is estimated that one out of twenty-five Caucasians—perhaps one child in every school classroom—is a CF carrier. How did the frequency of this apparently deleterious allele

get to be so high? There is some evidence to suggest that CF carriers, like the Hemoglobin-S heterozygotes we met in Chapter 2, may have an advantage over noncarriers under certain circumstances. It seems as though CF carriers may possibly have benefited from somewhat increased resistance to tuberculosis during the plague-ridden past of northern Europe. The same suggestion, by the way, has been tentatively made for carriers of Tay-Sachs, Gaucher's, and Niemann-Pick diseases, all common among people of Eastern European Jewish ancestry.

2. Unfortunately, understanding protein structure doesn't always lead to effective drug treatments; the specific alteration in Hemoglobin-S, for example, has been known for decades, but that knowledge has produced no treatment for sickle-cell anemia. On the other hand, work on the gene for fetal hemoglobin and its promoter may eventually pave the way to successful therapy.

3. Discussion about a national CF screening program, which would be the most ambitious effort of its type ever attempted, remains theoretical. That's because CF, unlike sickle-cell anemia, is not a unitary genetic condition: it can be caused by a host of different mutations. The most common CF defect, which accounts for 70 percent of U.S. cases, is the only one for which tests have been developed so far. More than eighty others have been found, and the list is growing. Any widely used test must pick up as many of those different defects as possible, or parents who have been assured that they do not carry a CF gene defect may in fact have one.

4. This program, instituted before the molecular revolution was in full swing, used a nongenetic method of screening, specifically, a solubility test for Hemoglobin-S. The issues raised by the experience, however, are applicable to genetically based testing as well.

5. The tests used most widely, like those developed earlier in the epidemic, do not test directly for HIV itself; they test for antibodies produced by the immune system in response to the virus. For our purposes here, "HIV test" refers broadly to tests designed, directly or indirectly, to indicate the presence or absence of HIV infection.

6. French Anderson, "Prospects for Human Gene Therapy," *Science,* vol. 226, pp. 401–409, 1984.

7. The ADA treatment was not the *very* first attempt at gene therapy, but it was the first one performed in the United States and the first to pass through the required review process. Earlier, a UCLA hematologist, Martin Cline, bypassed American restrictions and tried a rudimentary form of gene therapy on two thalassemia patients in Israel and Italy. The treatment had no effect, and Cline ended up losing most of his federal research grants. The ADA protocol opened doors for future trials because it survived the rigorous review—and, of course, because it succeeded.

CHAPTER 8

1. Constance Holden, "The Genetics of Personality," *Science,* vol. 237, pp. 598–600, 1987.

2. Stanley N. Wellborn, "How Genes Shape Personality," and John S. Lang, "Happiness is a Reunited Set of Twins," *U.S. News and World Report,* April 13, 1987, pp. 58–62.

3. Ibid. This particular article is not cited here as the most current opinions of either the article's authors or the magazine's editors. It is, however, typical of articles that have appeared and continue to appear in the popular press.

4. Kenneth Blum and Ernest Noble, "Allelic Association of Human Dopamine D2 Receptor Gene in Alcoholism," *Journal of the American Medical Association,* April 18, 1990.

5. Annabel M. Bolos, et al., "Population and Pedigree Studies Reveal a Lack of Association between the Dopamine D2 Receptor Gene and Alcoholism," *Journal of the American Medical Association,* December 26, 1990.

6. William S. Beck, Karel F. Liem, and George Gaylord Simpson, *Life: An Introduction to Biology* (New York: HarperCollins), pp. g–5.

7. Efforts to breed animals for specific behaviors don't always work out as planned. Before Gregor Mendel began working with garden peas, he tried breeding two races of honeybees to combine the best behaviors of both. His goal was to cross an industrious German breed of bees with a gentler, more attractive Italian strain. None of his hybrids, unfortunately, was hard-working, good-looking, or gentle.

8. Edward O. Wilson, *Sociobiology: The New Synthesis* (Cambridge, MA: Harvard University Press, 1975).

9. Ibid.

10. Ibid.

11. Eric S. Lander, "Splitting Schizophrenia," *Nature,* vol. 336, pp. 105–106, 1988. This situation, though still true today, may change in the not-too-distant future. Some researchers believe that hallucinations may become widely accepted as a central behavioral feature of schizophrenia, and others suggest that certain anatomical abnormalities in specific regions of the hippocampus may be shown to be common physical/physiological underpinnings.

12. We needn't look at behavior to find such situations. If, for example, we look for causes of the highly complex condition known as atherosclerosis, we find that a great deal more is involved than levels of blood cholesterol alone. Several genes in addition to the FH gene, for example, affect other risk factors, including levels of fats and other compounds in the blood, the ratio of HDL to LDL cholesterol, the tendency to develop high blood pressure, and a propensity toward obesity, to name just a few. Then there are the environmental factors involved: diet, exercise, cigarette smoking, stress, and even gout.

    As a result, although most people with high cholesterol are at increased risk of heart attack, not all of them are, because they may have an exceptionally favorable balance of other risk factors. Some people with only slightly elevated cholesterol levels can be at high risk of heart attack because those other risk factors are stacked in the wrong direction. Thus, there is still spirited debate, to use a popular euphemism, in the medical community concerning the contributing causes and the importance of cholesterol in the stew.

13. Thomas J. Bouchard, et al., "Sources of Human Psychological Differences: The Minnesota Study of Twins Reared Apart," *Science,* vol. 250, pp. 223–228, 1990.

14. Richard J. Herrnstein, "I.Q.," *Atlantic Monthly,* September, 1971, pp. 43–58.

15. John Aloysius Farrell, "Files Reveal Buchanan Views in the '70s," *Boston Globe,* January 4, 1992.

16. Farrell, *Boston Globe.*

17. Bruce Bower, "Gene in the Bottle," *Science News,* vol. 140, p. 191, September 21, 1991.

18. François Jacob, *The Possible and the Actual* (Seattle: University of Washington Press, 1982), pp. 65–66.

Acknowledgments

THE REAL SECRET of projects like *The Secret of Life* is that they depend on the hard work and goodwill of many people. This book has leaned on, caused sleepless nights for, or been blessed by the gracious advice of the following talented and dedicated individuals.

Paula Apsell, director of the WGBH Science Unit, initiated the project and committed the resources necessary to shepherd it through the maze of development, fund-raising, and staffing. Tom Levenson, former *Nova* science editor, shared with Joe Levine the challenge of assembling the National Science Foundation proposal that launched the series. As work commenced, Robin Brightwell, project director for the BBC, contributed a unique combination of wit, perspective, and production experience. We also thank series executive producer Graham Chedd for the singular atmosphere of camaraderie he fostered in the staff. This book would not exist without the efforts of Karen Johnson, director of publishing and merchandising at WGBH, and Moira Bucciarelli, who assisted with more behind-the-scenes details than we will ever know about. Nancy Lattanzio, project editor at WGBH, was an indefatigable, insightful, and compassionate guide to the delights and perils of publishing; her humor, empathy, and determination made her a delightful fellow traveler through the molecular world. Laurel Anderson did a superb job of tracking down elusive photos; only a fraction of her efforts is visible here. Thanks also to Virginia Jackson for help preparing the manuscript. Both the book and the series benefited immeasurably from the efforts of Susan Glass, whose official title of research assistant doesn't do her work justice. The rest of the team spoke "molecularese" as a second language at best; Susan is a native speaker whose fluency and insight opened doors for us all.

The series producers, a talented and creative group, each contributed greatly to the chapters dealing with their program material. In particular (and in order of presentation of their subject material), Linda Feferman, Andrew Liebman, Tim Haines, Chris Wizer, Paul Gasek, Larkin McPhee, and David Sington participated in discussions and working sessions that were personally and professionally rewarding. Assistant producers (in the same order) Janet Smith, Susan Laughlin, Bonnie Waltch, Elaine Jones, André Stark, Veronica Johow, Marisa Wolsky, and Joanna Timms helped coordinate vital details of program-book interaction. Carole Osterer, business manager for the series, was a constant reassuring presence.

As the manuscript took shape, fellow biologists participated in several ways. We owe a great deal to our central core of scientific advisers at the Whitehead Institute for Biomedical Research, especially Gerald Fink, Robert Weinberg, Harvey Lodish, Richard Mulligan, Eric Lander, and Richard Young. George Lyles of the Marine Biological Laboratory, Chris Simon of the Society for the Study of Evolution, and several individuals at the Rural Advancement Fund International, the National Wildlife Federation, and the Environmental Defense Fund helped us make contact with important researchers. Sources for the information we gathered were many and varied. In most instances, citations have been provided, but any quotations that are not footnoted are from the extensive interviews conducted by either the authors or the program producers. Individual chapters were improved and sharpened by the comments of eminent readers, including Robert Weinberg, Tim Tully, Eric Lander, Rob Dorit, Richard Young, Ronald Nagel, and Jackie Lederman.

Finally, Joe would like to thank Karel Liem, Edward F. MacNichol, Jr., E. O. Wilson, Richard Lewontin, and Evan Balaban, all of whom have shaped his perspective over the years in courses, discussions, and arguments on both the science and the social importance of biology. Much of what is correct in this volume owes a debt to these friends and colleagues.

The Secret of Life is a coproduction of WGBH/Boston and BBC-TV. Corporate funding for the television series was provided by the Upjohn Company, with additional funding from the National Science Foundation, Carnegie Corporation of New York, Alfred P. Sloan Foundation, Department of Energy, Lucille P. Markey Charitable Trust, Howard Hughes Medical Institute, National Center for Human Genome Research at the National Institutes of Health, the George D. Smith Fund, the Corporation for Public Broadcasting, and public television viewers.

Photo Credits

DNA extract: Jack M. Bostrack ©, Visuals Unlimited

Computer DNA: Jean Claude Revy ©, Phototake NYC

Sperm and egg: Francis Leroy, Biocosmos/Science Photo Library, Photo Researchers, Inc.

Human fetus: Petit Format/Nestle, Photo Researchers, Inc.

Maria Patino: Peter Freed/Alison Carlson

Yunnan, China: © 1988 John Eastcott/YVA Momatiuk, Woodfin Camp & Associates

Influenza: NIBSC/Science Photo Library, Photo Researchers, Inc.

Mammogram: © Chris Bjornberg, Photo Researchers, Inc.

Antibody: Courtesy of Genentech, Inc.

Senegal market: © Wolfgang Kähler

Sickle cells: © Al Lamme/Len/Phototake NYC

Tree snails: Chip Clark

Silversword: Gerald Carr

DNA injected into mouse embryo: © Jon Gordon/Phototake NYC

Fermentation tank: © Michael Grecco, Picture Group

Transgenic ewes: Pharmaceutical Proteins Ltd.

Boy and chimp: © Philip Herman

Twins: Dembinsky Photo Assoc.

Eugenics Journal: Sophia Smith Collection, Smith College, Northampton, MA

Index

A

B

C

H

I

J

K

L

M

N

O

P

Q

R

S

T

U

V